1st Workshop on Metrology for Agriculture and Forestry 2018

IOP Conference Series: Earth and Environmental Science
Volume 275

Ancona, Italy
1 - 2 October 2018

Editor:

Enrico Primo Tomasini

ISBN: 978-1-5108-8746-6
ISSN: 1755-1307

Printed from e-media with permission by:

Curran Associates, Inc.
57 Morehouse Lane
Red Hook, NY 12571

Some format issues inherent in the e-media version may also appear in this print version.

Institute of Physics
Dirac House, Temple Back
Bristol BS1 6BE UK

Phone: 44 1 17 929 7481
Fax: 44 1 17 920 0979

techtracking@iop.org

Additional copies of this publication are available from:

Curran Associates, Inc.
57 Morehouse Lane
Red Hook, NY 12571 USA
Phone: 845-758-0400
Fax: 845-758-2633
Email: curran@proceedings.com
Web: www.proceedings.com

TABLE OF CONTENTS

Proceedings of the 1st Intl Workshop on Metrology for Agriculture and Forestry
Ancona, Italy, 1-2 October 2018
https://www.metroagrifor.org/maf2018/

The **1st Intl Workshop on Metrology for Agriculture and Forestry** was organised by the Departments of Industrial Engineering and Mathematical Sciences and Agricultural, Food, and Environmental Sciences of Università Politecnica delle Marche, Italy, in collaboration with A.I.S.S.A.- the Italian Association of Scientific Agriculture Societies with the support of A.I.VE.LA. – the Italian Association of Laser Velocimetry and Non-invasive diagnostics.

The event, which took place at the premises of the Faculty of Engineering in Ancona on 1-2 October 2018, aimed at creating an active and stimulating forum where academics, researchers and industry experts in the field of measurement and data processing techniques for Agriculture, Forestry and Food could meet and share new advances and research results.

The Workshop featured invited lectures, presentations, poster sessions as well as a **Short Course on Measurement and Uncertainty**, designed to provide the basics of measurement, calibration and uncertainty evaluation, and an **Exhibition**, to offer leading manufacturers the opportunity to bring their latest technology solutions directly to the attention of a selected audience.

This volume contains a selection of the papers presented at the event.

The detailed Programme of the Workshop can be found at: https://www.metroagrifor.org/maf2018/programme/.

I trust the initiative will meet your interest and thank you in advance for your kind attention.

EDITOR
Prof. Enrico Primo Tomasini
Department of Industrial Engineering and Mathematical Sciences
Università Politecnica delle Marche
Ancona, Italy

Peer review statement

All papers published in this volume of *IOP Conference Series: Earth and Environmental Science* have been peer reviewed through processes administered by the proceedings Editors. Reviews were conducted by expert referees to the professional and scientific standards expected of a proceedings journal published by IOP Publishing.

IOP Conf. Series: Earth and Environmental Science **275** (2019) 012001 doi:10.1088/1755-1315/275/1/012001

Challenges of multi/hyper spectral images in precision agriculture applications

Adriano Mancini, Emanuele Frontoni, Primo Zingaretti

VRAI Lab, Dipartimento di Ingegneria Informatica, Università Politenica delle Marche, Ancona, ITALY

E-mail: a.mancini@univpm.it, e.frontoni@univpm.it, p.zingaretti@univpm.it

Abstract. According to the CEMA, Agriculture 4.0 paves the way for the next evolution of farming consisting of unmanned operations and autonomous decision systems while Agriculture 5.0 will be based around robotics and (some form of) artificial intelligence. It is clear how Agriculture is experimenting a Copernican Revolution where multidisciplinarity is the engine of this revolution. In this context it is central the topic of Variable Rate Treatments (VRTs). A VRT application relies on prescription maps that are generated by considering the agronomist experience augmented with data sensed also by using advanced platform as Unmanned Aerial Vehicles (UAVs) or satellites. Prescription maps are usually implemented by tractors using Variable Rate Controllers (VRCs) that apply a given quantity of product on a given region. The typical operation is spraying nitrogen or weed treatment application over the agricultural field. The generation of a correct prescription map requires the definition of specific management zones that reflect areas and their status. The planning of agricultural tasks necessitates a deep knowledge of crop state, for example, an important but typical case is variable rate nitrogen fertilizer application. In this scenario multi-spectral images play a key role and today the technology is mature to be used also in real applications. Hyper-spectral imagery is still expensive and it is usually adopted for advanced research considering the complexity to acquire and post-process data. However multi and hyper spectral systems are changing the agriculture enabling complex analysis with ultra high spatial and spectral resolution.

1. Introduction

Over the last few years, agriculture is experiencing a sort of "Copernican revolution" dictated by the use of high-performance technologies also supported by artificial intelligence. This is also underlined in the report of the European Committee of Agricultural Machinery Manufacturers' Groups, which states that Agriculture 5.0 will be based on the use of robotic systems and artificial intelligence [1]. In this scenario, there is an ever increasing agrifood literacy of producers and consumers with references to the traceability of the supply chain and its environmental sustainability. It is clear that precision and sustainability are issues that are essential today. Precision consists in the optimal use of a resource with the objective of maximizing yield, minimizing costs and environmental impact. This complex optimization can be made possible if and only if man, machine and robots [2] work synergistically in measuring, evaluating, deciding and implementing prescriptions in a continuous cycle. This today can be achieved through multi-sensorial systems that operate at different spatial scales (where), temporal (when) and spectral (what) [3]. The centrality lies therefore in the data that by its nature in a scenario of Agriculture 4.0 / 5.0 becomes complex to manage both in terms of size and complexity of the

IOP Conf. Series: Earth and Environmental Science **275** (2019) 012001 doi:10.1088/1755-1315/275/1/012001

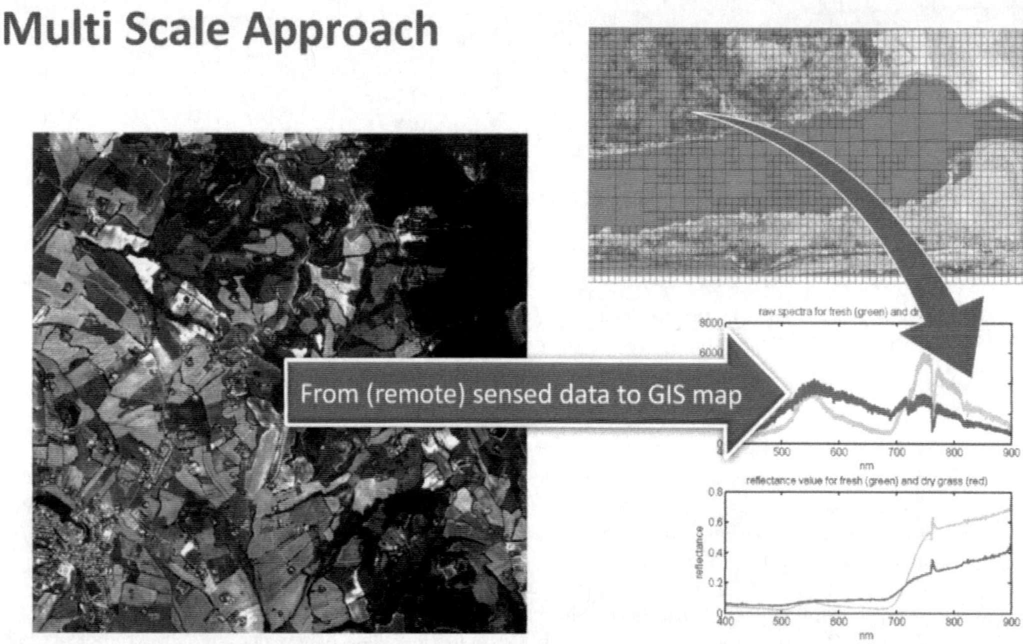

Figure 1. Processing of hyper/multi-spectral acquisition should consider different spatial, spectral and temporal scales considering also the complexity related to the data collection.

analysis to be carried out. An example is the ability to identify crop vs soil that in some cases can be very complex due to *grassing* [4] techniques.

2. Main Challenges of Multi/Hyper Spectral Data

During last years multi-spectral and hyper-spectral sensors have been applied in Precision Agriculture applications [5]. The state-of-the-art technologies today acquire a hundreds of bands in the VISible (VIS), Near InfraRed (NIR) and Short Wave InfraRed (SWIR). Actually it is possible to mount these sensors also on ground/aerial (unmanned) vehicles to precisely map fields and crops by identifying stress as nitrogen or water stress to derive custom and precise prescription maps that modern tractors are able to execute by using the Variable Rate Controller technology [3].

The high spatial and temporal resolutions require new approaches to process this huge amount of data [6]. It is also important to consider geometric and radiometric corrections in order to have useful data [7]. Radiometric correction plays a key role and several techniques area available with different pros and cons as radiometric target or real time incident light sensors [8]. Moreover ground and aerial sensors could be supported by satellite technologies that enable in some case weekly or daily updates [9, 10]. We performed several flights to collect multi/hyper-spectral cubes also developing new approaches to process data that include ground spectrometer and aerial sensors on different crops with different nitrogen and irrigation configuration in cooperation with the D3A Department of Universit Politecnica delle Marche. Figure 1 shows the approach that includes different spatial, spectral and temporal scales.

Today Satellites provide images with Ground Sampling Distance (GSD) that typically ranges from 0.5m to 1000m with a number of bands from 3 to 15-20 (in case of multispectral payloads) and some platform provides a worldwide coverage every 1-2 days. In this scenario precision agriculture gains the capability to monitor the trend of some vegetation index over time acting as a scouting service. In case of potential issues (i.e. area with a anomalous values of vegetation index) it could be possible to perform an acquisition by using an unmanned platforms that usually generate ultra high resolution images [11]. Satellites must be integrated by unmanned platforms [12] [13] in order to increase the spatial and also the spectral resolution. The use of unmanned vehicles as multi-rotor and fixed wing is quite common in precision agriculture [14, 15] even if the endurance and then the coverage are the main bottleneck. In particular hyper-spectral systems require low flight speed at altitude lower than 100m to avoid blurred images also considering that exposure time changes according to the detector technology. A complete field survey includes satellite and unmanned platform and also it is recommended to acquire ground data by using spectrometer [16]. The merging of data is a complex matter considering that each sensing platform has different techniques to provide calibrated data also considering that spectral configuration is not the same for all devices.

2.1. Calibration

The radiometric calibration is one of the most critical phase during the acquisition of multi/hyper-spectral images. This process enables the conversion from the Digital Number (DN) to a reflectance value. Recently this activity have been deeply studied with particular reference to Reflectance Mode Method, Linear-interpolation Method, Continuous Panel Method, Multi-band radiometer method [17, 18]. The calibration plays a key role to set-up the data processing in the right way [19, 20] also aided by in-situ calibration [21] for long term monitoring. If calibration is not properly performed then wrong results and then decisions could be generated as discussed in [22]. Moreover it is important to compensate changes in illumination mainly caused by clouds. In this case it is necessary to use incident light sensor to have properly calibrated data [23].

2.2. High resolution and noise reduction

Hyper-spectral linescan sensors require a complex post-processing to output co-registered hyper-spectral cubes. We developed algorithms to automate the detection of ground targets with well-known position and reflectance in order to facilitate the co-registration tasks since the sensor used was not equipped, for reasons of weight and cost, with a high precision and accuracy INS and GPS RTK. For each group of pixels and for each band we extract an histogram. Starting from this we isolate the response of the soil and the crop using the very high spatial resolution that reaches 1-2cm. Figure 2 shows an example of a histogram for a group of pixels for a given band. Systems have also been developed to perform a scouting by means of spectrometers mounted on board o low cost multi-rotor unmanned platform characterized by a very low spatial resolution, but at the same time with a very high spatial resolution [7].

From the analysis of the histograms for a given set of geographical areas that follow a given experimental scheme to induce *nitrogen* and *water stress* it is possible to extract complex vegetation indices that highlight the differences by extending indices such as NDVI and NDRE. The definition of these indices can then be used to identify the presence or absence of *stress* using data acquired with low-cost multi-spectral cameras with bands selected from the analysis of hyper-spectral data. We started from previous methods developed in the Land Use / Land Cover (LU / LC) presented in [10].

2.3. Soil vs Crop Segmentation

High and ultra-high resolution images (hyper or multi spectral) should be post-processed in order to remove the effect of soil that could generate wrong results during the data analysis. In this case it is necessary to mask out soil areas in order to evaluate only vegetation.

Typically the soil/crop segmentation is performed by using *thresholding* algorithms starting from vegetation indices such as Normalized Difference Vegetation Index (NDVI) that have

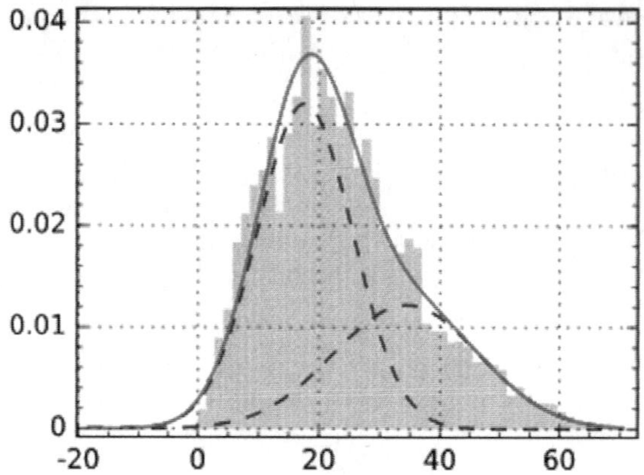

Figure 2. The histogram represents the reflectance dynamics (x-axis) for a group of pixels for a given band. It is possible to distinguish two contributions due in this example from soil and crop.

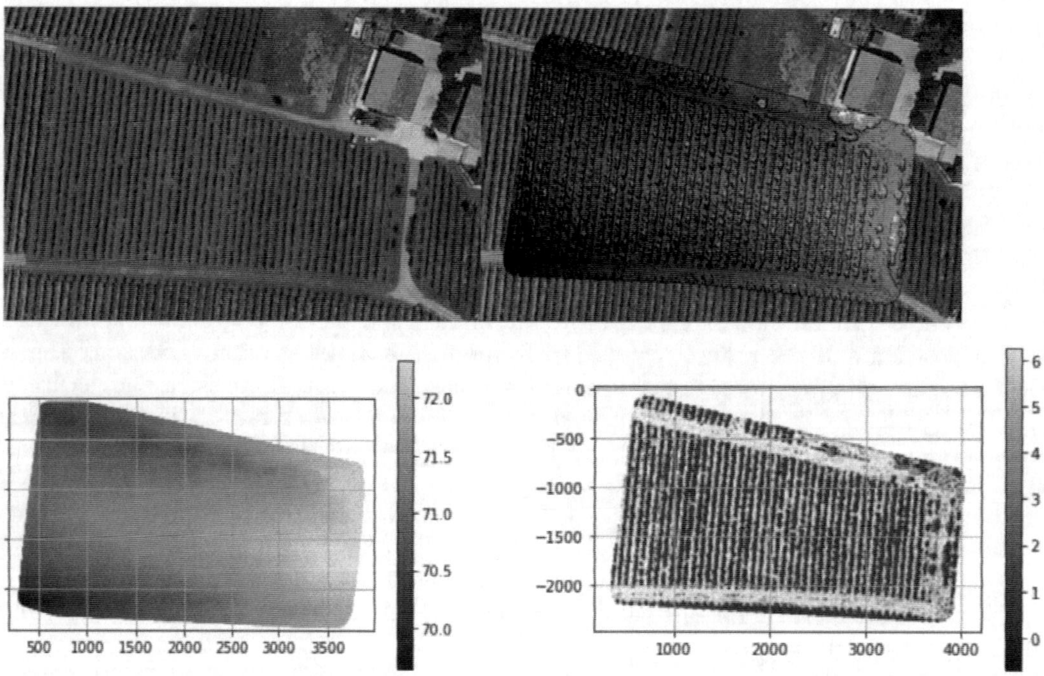

Figure 3. top-left: false colour image; top-right: DSM; bottom-left: extracted soil with algorithms as in [4]; bottom-right: extracted vegetation.

problems in *grassland* situations. In this situation it is necessary to use the Digital Surface Model (DSM) model in order to correctly discriminate between *soil* and *crop* in contexts such as vineyards. There have been several approaches such as CARSCAN and FANSCAN [4] that start from the DSM and then integrates the radiometric response if necessary to improve the overall quality (e.g. Figure 3).

References

[1] CEMA 2017 Digital farming: what does it really mean? Tech. rep.
[2] Duckett T, Pearson S, Blackmore S and Grieve B 2018 *arXiv preprint arXiv:1806.06762*
[3] Mancini A, Frontoni E and Zingaretti P 2018 *2018 International Conference on Unmanned Aircraft Systems, ICUAS 2018* 856–863
[4] Mancini A, Dyson J, Frontoni E and Zingaretti P 2017 *2017 International Conference on Unmanned Aircraft Systems, ICUAS 2017* 1410–1417
[5] Zhang C and Kovacs J 2012 *Precision Agriculture* **13** 693–712
[6] Sturari M, Frontoni E, Pierdicca R, Mancini A, Malinverni E, Tassetti A and Zingaretti P 2017 *European Journal of Remote Sensing* **50**
[7] Mancini A, Frontoni E and Zingaretti P 2016 *2016 International Conference on Unmanned Aircraft Systems, ICUAS 2016* 1148–1155
[8] Mancini A, Frontoni E, Zingaretti P and Longhi S 2015 *2015 International Conference on Unmanned Aircraft Systems, ICUAS 2015* 534–542
[9] Bernardini A, Frontoni E, Malinverni E, Mancini A, Tassetti A and Zingaretti P 2010 *Journal of Spatial Science* **55** 43–54
[10] Malinverni E, Tassetti A, Mancini A, Zingaretti P, Frontoni E and Bernardini A 2011 *International Journal of Geographical Information Science* **25** 1025–1043
[11] Bansod B, Singh R, Thakur R and Singhal G 2017 *Journal of Agriculture and Environment for International Development* **111** 383–407
[12] McCabe M, Houborg R and Lucieer A 2016 *Proceedings of SPIE - The International Society for Optical Engineering* **9998**
[13] Abdullahi H, Mahieddine F and Sheriff R 2015 *Lecture Notes of the Institute for Computer Sciences, Social-Informatics and Telecommunications Engineering, LNICST* **154** 388–400
[14] Adao T, Hruka J, Pdua L, Bessa J, Peres E, Morais R and Sousa J 2017 *Remote Sensing* **9**
[15] Rudd J, Roberson G and Classen J 2017 *2017 ASABE Annual International Meeting*
[16] Cozzolino D 2014 *Applied Spectroscopy Reviews* **49** 564–584
[17] Peddle D R, White H P, Soffer R J, Miller J R and LeDrew E F 2001 *Computers & Geosciences* **27** 203 – 213 ISSN 0098-3004
[18] Miura T and Huete A R 2009 *Sensors* **9** 794
[19] Burkart A, Aasen H, Alonso L, Menz G, Bareth G and Rascher U 2015 *Remote Sensing* **7** 725 ISSN 2072-4292 URL `http://www.mdpi.com/2072-4292/7/1/725`
[20] Zhao F, Li Y, Dai X, Verhoef W and Guo Y 2015 *Remote Sensing of Environment* **156** 129 – 142 ISSN 0034-4257
[21] Jin H and Eklundh L 2015 *IEEE Transactions on Geoscience and Remote Sensing* **53** 3405–3416
[22] Hruska R, Mitchell J, Anderson M and Glenn N F 2012 *Remote Sensing* **4** 2736 ISSN 2072-4292
[23] Hunt ER J and Daughtry C 2018 *International Journal of Remote Sensing* **39** 5345–5376

IOP Conf. Series: Earth and Environmental Science **275** (2019) 012002 doi:10.1088/1755-1315/275/1/012002

Comparison of 3D Imaging Technologies for Wheat Phenotyping

Imran Mohamed and Richard Dudley

National Physical Laboratory, Hampton Road, Teddington, Middlesex, TW11 0LW, United Kingdom

Email: imran.mohamed@npl.co.uk

Abstract The in-field measurement of phenotypes or traits of wheat such as ear size is important data for use in the development of newer wheat varieties. The data is currently gathered manually from hundreds of test plots by random sampling of the what within each plot. To improve the data quality and data collection speed, we investigate and compare the use three different 3D imaging technologies: multistereo imaging, time-of-flight and structured light laser scanning to produce point clouds of a wheat plant in-situ. Measurements of the wheat plant's ear is made from the generated point clouds.

1. Introduction

Plant phenotyping is the quantification of a plant's physical traits, genetics and ontogeny. For wheat, key phenotypes include yield, ear size and disease resistance [1]. Phenotype data is used to guide cross-breeding programmes for new varieties and an efficient process requires large phenotype data sets, recorded in a quantifiable and accurate way [2]. Phenotype data ideally should take place in-situ on outdoor plots with variations in locations, soil type, drainage etc.

Capturing data for just wheat ear size, currently relies on manual sampling with a ruler [3]. Measurements are repeated on hundreds of plots in the same field, each growing different wheat varieties. Data collection is time-consuming, and the quality of the data limited by the sample size within each plot, creating a data bottleneck in capacity. 3D imaging of field grown crops plots expands the quantity of data that can be collected, is faster than making measurement manually and enables more post processing for additional plants trait information. Measurement data on ear size is obtained by shape fitting to 3D point clouds, a virtual representation of a scene, created by commercial imaging hardware [4] and software [5][6].

In this paper we present trials of the three of the most promising imaging technologies for field phenotyping deployment, multi-stereo imaging, Time of Flight cameras (ToF) and Structured Light Laser Scanning (SLLS). All the technologies have been demonstrated for indoor plaint trait analysis [7], [8] but some of their application outdoors is less well reported.

2. Imaging Technologies

Multi-stereo imaging uses multiple images of a scene taken from multiple positions. Using stereo matching algorithms, the images are analysed with information about the cameras' positions and internal parameters such as focal length. Pixels in captured images are assigned an x,y,z location using stereo-matching algorithms resulting in a point cloud representation of the image. Stereo matching algorithms include those found in commercial software such as HALCON [5] or open source libraries such as Point Cloud Library [6]. In this study the multigrid stereo matching algorithm implemented in HALCON was utilised.

A time-of-flight camera is more comparable with radar than conventional cameras. TOFs use pulses of near-infrared (NIR) light to illuminate a scene, with pulse lengths in the order of 10 ns. A high-speed CCD records the reflected NIR images and the time of arrival. and a distance can be calculated. Each pixel on the CCD sensors thus records intensity and distance enabling a point cloud of the scene to be produced in real time at typically 20 fps and with a resolution of 640×480 pixels.

SLLS is the latest variation on laser projection scanning and uses a moving pattern of stripes projected onto a scene using a high-brightness laser. A CCD camera captures the changing patterns on the scene and converts the distortions around surfaces directly into a point cloud. From the observed distortions it is possible to calculate the 3D structure of the scene and convert this data into a point cloud.

3. Image Capture

Obtaining 3D images of wheat outdoors is challenging. Variable lighting conditions and wind influence the quality of the image and/or the ability of the instrument to acquire an image. At NPL we constructed a 2 x 2 m wheat trial plot enclosed by two linear motion rails and two A-frame camera mounts. The rails were placed on the north and south side of the plot to allow the A-frames to move in an east-west direction, which we define as the x-axis, Figure 1.

Figure 1: Wheat plot and imaging platform as viewed from its southern side.

Our study initially focussed on imaging a single wheat plant to validate and calibrate imager operation. The wheat plant was chosen due to its location in the less densely populated end of the plot which allowed easier imaging without occlusion from neighbouring wheat plants. The imaging instruments were mounted onto the A-frame and images captured during the afternoon, with ambient light levels varying between 3000 to 4000 lux. CloudCompare [9] was used to view and post-process measurements from the produced point clouds.

For multi-stereo imaging, four 2.3-megapixel monochrome Basler cameras were used to capture images of the wheat plant. Prior to mounting on to the frame, the cameras' internal parameters were calculated using the calibration routine in HALCON. The routine involves the use of a 32 cm × 24 cm calibration plate consisting of 5 mm white circles arranged hexagonally on a black background, with the centre to centre separation of each circle being 1 cm. Images of the calibration plate placed in a variety of positions and orientations are taken. By comparing the size and position of the dots seen across in the camera's field of view from what is expected the software calculates the: focal length, radial lens distortion parameters, decentring lens distortions, the centre of the radial lens distortion, and sensor's physical dimension.

The same calibration plate used for camera calibration was placed on the southern side of the wheat and was used to calculate the external camera parameters of position and orientation. Figure 2 shows the imaging set up during assembly of the experiment with one pair of monochrome cameras mounted to the southern A-frame. The remaining pair of monochrome cameras were mounted on the northern A-frame. The lens to lens centre distance between each camera pair was 10 cm. The cameras were positioned so that all four were able to see the wheat plant and calibration plate.

Figure 2: Imaging setup with pair of monochrome cameras and ToF camera attached to A-frame on southern side (left in the figure) of plot. Also visible is the calibration plate used for the multistereo imaging placed on the southern side of the wheat plant being imaged.

With an exposure time of 5 ms, images from the four cameras were captured, then the two A-frames moved approximately 5 cm and capture repeated. This was carried out for as long as the calibration plate was visible from all the four cameras, example frames are shown in Figure 3. The captured images were processed using the multigrid stereo matching algorithm implemented in HALCON to produce a point cloud.

Figure 3: The view of the wheat plant under investigation and calibration plate from the four monochrome cameras.

The larger leaves and stems structures were recreated well by the algorithm as seen in Figure 4. However, wheat ears often were not well-formed image focus, crop movements and algorithm pixel-correlation challenges within the captured images. The greater the number of input images taken from different positions and orientations improve the quality point cloud at the expense of computational and capture time. The point cloud of Figure also contains significant noise from structures that were not fully formed by the algorithm. Simple denoising techniques such as CloudCompare's Statistical Outlier Removal filter can improve the final quality of the point cloud but does at this stage require manual optimisation.

Figure 4: Multistereo imaging point cloud.

For Time of Flight trials, a Basler ToF camera was mounted on the southern A-frame at the same time as the monochrome camera pair shown in Figure 2. The ToF camera's default processing mode, High Dynamic Range (HDR), was used which allows the camera to take both short (4 ms) and long (20 ms) exposures and combine them into a single image. This allows the camera to achieve a more even NIR illumination across the scene which in turn allows for more accurate distance measurements.

The ToF camera produced the point cloud shown in Figure 5 with the structure of the wheat plant clearly visible for most of the stems and leaves, but not all the ears are visible. In contrast the larger structures around the wheat plant such as the calibration plate, the imaging frame and the ground itself are visible highlighting the ToF cameras low resolution and inability to resolve the smaller structures of the wheat plant.

Figure 5: The Time of Flight point cloud.

The final imaging technology used a structured light laser scanner (SLLS), the Photoneo PhoXi 3D. Due to its larger size it was mounted separately from the monochrome and ToF cameras. The scanners longer acquisition time of 1 to 2 seconds, compared to the 10 ms of the monochrome and ToF cameras, meant more care was required to ensure image capture was taken when the wheat was stationary.

The structured light laser scanner, having a higher resolution, was able to produce point clouds of the wheat, Figure 6, including stem, leaves, ears and even florets. Less of the surroundings are seen in the point cloud due to the smaller scanning volume.

Figure 6: Structured light laser scanning point cloud.

4. Imaging Technology Comparisons

Comparing the three imaging technologies, the SLLS produced the best quality point cloud out of the three presented technologies. It captured the stems, leaves and ears in high resolution and minimal noise. Multistereo imaging was next best due to it also being able to recreate most of the wheat plant's structure, albeit with noise. ToF performed the worst, only being able to discern the main structure but without enough detail to be of use.

Based on these initial results, multistereo imaging's advantages compared to ToF and SLLS are its ability to produce point clouds at low cost (each monochrome camera cost in the order of GBP 100) and function under direct sunlight. An ability to obtain images with exposure times in the order of hundreds of milliseconds means motion blur caused by wind can be minimised without the use of wind shield. Being able to "freeze" any motion in this way will also be advantages should the technique need to be used from a moving platform. Disadvantage of multistereo imaging is the computing time and resources required to process the input images into a point cloud. The presented point cloud required over 20 seconds for the stereo matching algorithm to complete on an Intel i7-7500U processor, reaching a peak RAM usage of 1.7 GB. This would increase as the imaging volume increased (e.g. imaging a row of wheat plants versus a single wheat plant) and the number of input images (e.g. to capture more wheat plants or to improve the point cloud of a single wheat plant). Setup time was lengthier than ToF and SLLS due to the number of monochrome cameras that required mounting on to the imaging frame.

The ToF camera, despite not producing a usable point cloud, has the advantage that it can produce a point cloud in real-time at up to 20 frames per second rather than requiring offline processing of images. Exposure times can be reduced to 300 μs which allows for mitigation against motion blur of the wheat plants caused by light wind although the monochrome cameras can reach exposure times as low as 40 μs which allow for even less motion blur provided there is sufficient lighting. Its cost is in the order of GBP 1000. The two primary disadvantages are the camera's low resolution which impeded it from resolving as much wheat structure as the other technologies achieved and its inability to work under strong or direct sunlight (the camera specifications state a maximum ambient light level of 15 000 lux). In future uses these effects will be mitigated by moving the ToF camera closer to the scan subject and using a screen to block excess ambient light.

The advantage of the SLLS is the high quality of the produced point cloud, short setup time and being able to produce a point cloud directly. The disadvantages to using the SLLS are its high cost (over GBP 10 000), an inability to work under direct sunlight and acquisition times in the order of seconds which make it susceptible to wheat movement caused by wind. A screen could be used to limit the effects of sunlight, as can performing the scan at night or when light levels are low. The screen would also mitigate effects of wind during the seconds long image capture, but only if the SLLS is stationary. Using SLLS to phenotype large fields from a continuously moving imaging platform would not work.

4.1. Ear Height Measurement

Measurements were only carried out using the SLLS produced point cloud because it was the only one where almost all the wheat ear's height was recreated. Manual measurement on the wheat plant of interest was carried out with a tape measure and a virtual point cloud measurement was carried out using CloudCompare (Figure 7). The manual measurement on the left wheat ear in Figure 7 was 83 mm in height from ear base to the top of the spike. The point cloud measurement taken from the lowest and highest visible wheat ear points came to 82.4 mm. The right wheat ear in Figure 7 came to 86.7 mm when measured manually and 81.7 mm when measured from the point cloud, a lower value caused by the bottom most floret and the stem not being imaged in the point cloud leading to an under measurement.

Figure 7: Measurement of the wheat ear of interest in the field (left) and with the SLLS produced point cloud (right).

This demonstrates that measurements of wheat ear height carried out with point clouds can be as accurate as those obtained manually in the field if the point cloud closely reproduces the structure of the wheat ear. To take the method further, more than a single wheat plant needs to be imaged and measured. With greater number of wheat ears in a point cloud this also necessitates a way to automate the detection and measurement of wheat ears in a point cloud. An expanded measurement campaign is planned for the next growing season with the intent of producing point cloud based measurements of more of the plot.

5. Conclusions

An investigation into the use of multistereo imaging, ToF and SLLS to produce point clouds for in-field dimensional phenotyping was completed. The preliminary results based on the imaging of a single wheat plant show that multistereo imaging and SLLS are both capable of producing point clouds that show the structure of the wheat plant. However only the SLLS point cloud was detailed enough to obtain a measurement of the ear height. Due to the SLLS high cost and the advantages possible with it, suggestions on how improvements to the point cloud produced via multistereo imaging were given.

Acknowledgements
This work is supported by the UK government's Department of Business, Energy and Industrial Strategy (BEIS).

References

[1] Walter A, Liebisch F and Hund A 2015 Plant phenotyping: from bean weighing to image analysis *Plant Methods* **11** 14
[2] Aruas J L and Cairns J E 2014 Field high-throughput phenotyping: the new crop breeding frontier *Trends Plant Sci.* **19** 52–61
[3] Torres A and Pietragalla J 2012 Crop morphological traits *Physiological breeding ii: a field guide to wheat phenotyping* eds A Pask *et al* (Mexico, D.F.: CIMMYT) p 106
[4] Photoneo. https://www.photoneo.com/
[5] HALCON. MVTec Software GmbH
[6] Rusu R B and Cousins S 2011 3D is here: Point Cloud Library (PCL) *IEEE ICRA*

[7] Rose J C, Paulus S and Kuhlmann H 2015 Accuracy analysis of a multi-view stereo approach for phenotyping of tomato plants at the organ level *Sensors* **15** 9651–65

[8] Wang Y, Wen W, Wu S, Wang C, Yu Z, Guo X and Zhao C 2019 Maize plant phenotyping: comparing 3D laser scanning, multi-view stereo reconstruction, and 3D digitizing estimates *Remote Sens.* **11** 63

[9] CloudCompare. http://www.cloudcompare.org/

IOP Conf. Series: Earth and Environmental Science **275** (2019) 012003 doi:10.1088/1755-1315/275/1/012003

Strawberry ripeness calibrated 2D colour lookup table for field-deployable computer vision

Imran Mohamed, David Williams, Richard Stevens and Richard Dudley

National Physical Laboratory, Teddington, Middlesex TW11 0LW, United Kingdom

Email: imran.mohamed@npl.co.uk

Abstract. Efficient strawberry harvesting using semi-autonomous robotic pickers requires a method for fast, automatic discrimination between ripe and unripe berries. In this paper we present a strawberry ripeness algorithm that has been developed and tested on images obtained with commercial cameras, recorded under daylight illumination on an outdoor strawberry farm. The method uses a 2D lookup table seeded such that regions for ripe strawberry, unripe strawberry and strawberry calyx exist in distinct locations in the table. Measurements show the ratio of ripe to unripe area provides a good estimator of ripeness and that the approach was practical for further extension to real-time use.

1. Introduction

In recent years the post-harvest processing of fruit has become a highly automated process reliant on technology rather than manual labour. The use of cameras and machine vision has improved the speed and efficiency with which processes such as grading and sorting are carried out [1-3], reducing costs to the producer and consumer. Fruit harvesting on the other hand remains a largely manual process, but progress is being made on automation for a variety of fruits [4].

Strawberries, with a high value and high labour costs are particularly attractive for semi- and fully-automated harvesting solutions, particularly as in the UK, labour shortages are becoming increasingly more challenging [5].

Once picked, strawberries cease ripening [6] and have short shelf lives, so it is important to pick them at the right time to maximise acceptance by the consumer. Therefore, fast and accurate computer vision algorithms are required to quantify strawberry ripeness for a pick decision. Unripe strawberries are green-white in colour, due to the presence of chlorophyll in their skin. As ripening progresses, patches of red, caused by anthocyanin, emerge and increase in surface area [6]. Therefore, an obvious way to define a strawberry's ripeness from a computer vision perspective is to ratio the amounts of ripe and unripe surface area to produce a ripeness factor.

Determining which areas are ripe and unripe requires colorimetry [7], quantification of spectral reflectance across the visible spectrum. A spectrometer could be used to measure the reflectance spectrum, from which an accurate determination of colour can be calculated [8]. However, due to the measurements being taken with a fibre optic device, acting effectively as a single pixel, only a limited area can be inspected at a time. Scaling this up to provide multiple readings from each strawberry would slow down a harvester. Hyperspectral cameras, on the other hand, would allow images to be captured much like conventional cameras, but also take spectral readings for every pixel in the image with a resulting significantly higher cost than a conventional camera. Spectrometers and hyperspectral cameras could provide the necessary spectral measurements, but as a harvesting robot will already deploy single or multiple conventional cameras for berry location, a method to utilise these same cameras is preferential. Conventional cameras are much lower cost than spectrometers and hyperspectral cameras, but spectral information is not directly provided.

In this paper we present a strawberry ripeness estimation method that uses a lookup table to segment the ripe and unripe areas from an RGB camera image and is suitable for deployable computer vision systems. It was designed as part of the AUTOPIC project to develop a prototype autonomous robotic strawberry picker. As part of the robot's modular functionality, the ripeness estimation module was designed and ran separately from the strawberry detection module. The ripeness estimation module was also required to allow manual setting and intervention of the range of acceptable ripenesses that the robot picked. The paper presents an overview of the approach, lookup table structure and utilisation on test images obtained at an outdoor, commercial strawberry farm. Unlike other strawberry ripeness evaluation methods [9-12] our tests are undertaken in an outdoor environment rather than laboratory or on enclosed conveyer belts conditions.

2. Lookup table method

2.1. Chromaticity coordinates

Commercial colour cameras provide three component values at each pixel forming the image: red, green and blue (R, G, B). The colour seen by an observer depends primarily on the relative values of these components, and to a much smaller extent on their absolute values, whose sum relates to the overall intensity at the image point. It is difficult to achieve a good accuracy in correcting the observed intensity for prevailing lighting conditions [8]. It is therefore convenient to normalise the three values by dividing by their sum, giving $R/(R+G+B)$ etc., which eliminates the overall intensity and yields three "chromaticity coordinates" (r, g, b) whose total is unity for the red, green and blue component values. The colour can then be defined by just two of the three numbers. Given the red and green coordinates, blue is implicit as the difference from unity.

2.2. 2D colour lookup table

To segment an image of a strawberry into different regions based on colour, one may define an acceptable range of values of each chromaticity coordinate for every defined region requiring segmentation. These ranges are then applied to every pixel in the input image to check whether it meets the requirements for any of the segmented regions. An alternative approach to this is to use a colour lookup table. This is an arrangement of all possible combinations of colours reproducible using the chromaticity coordinates assigned to the table's dimensions.

Using the red and green chromaticity coordinates, r and g, a basic two-dimensional table with $n \times n$ elements can have dimensions K and L defined as $K = nr$ and $L = ng$ respectively. For this work a value of $n = 255$ was chosen due to standard commercial cameras producing 24-bit colour images with each colour having integer values between 0 and 255. When appropriate K and L are rounded to the nearest integer values. The resulting table, illustrated in figure 1, has the colours distributed in a triangular region covering only half of the table's available elements, since $K + L$ has a maximum value of n. As a coordinate (K, L) may be obtained by more than one combination of R, G and B, corresponding to differences in intensity, Figure 1 shows the colours with R, G and B scaled so that their values are as large as possible.

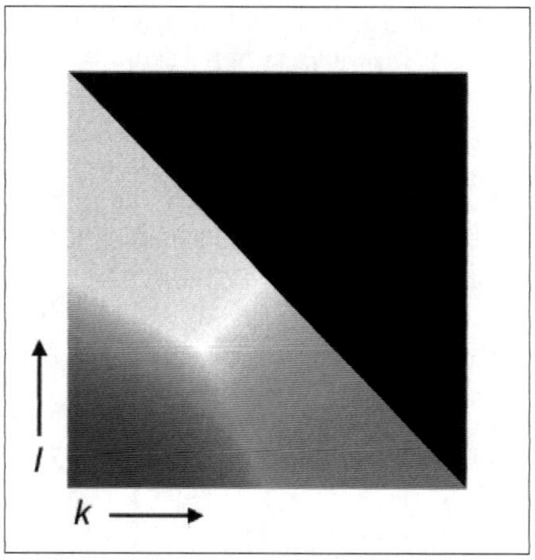

Figure 1. Basic 2D colour lookup table.

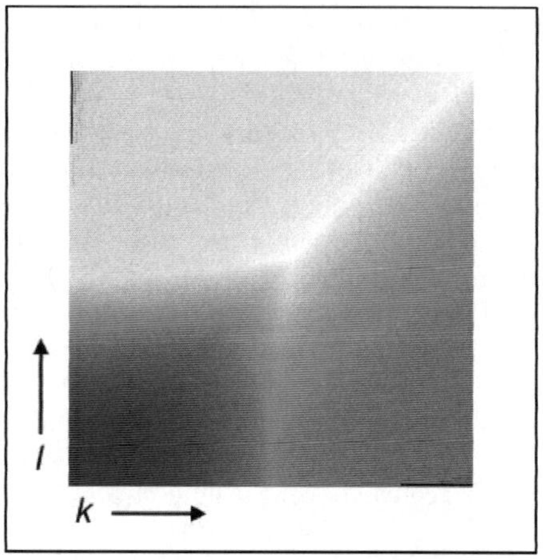

Figure 2. Modified colour lookup table.

2.3. Modified lookup table

It would be more acceptable to an operator to make use of the full table by removing the blank region. Pictorially this can be done by stretching the yellow midpoint of the triangle's green-red edge into the lookup table's vacant corner, as in figure 2. The resulting arrangement of colours is like that of the Natural Colour System [13], in which a fourth basic colour, yellow, is introduced to the standard red, green and blue.

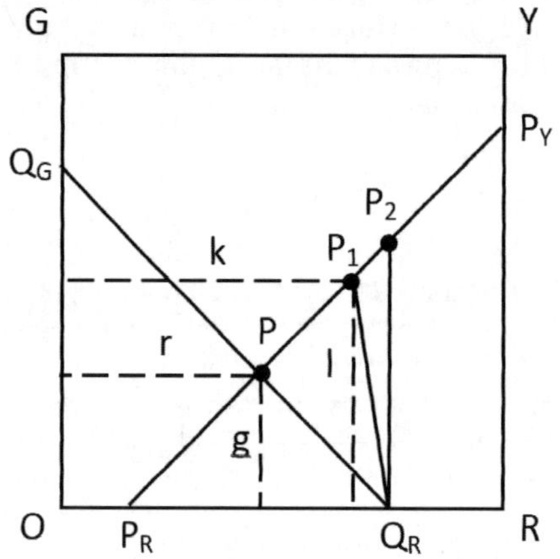

Figure 3. Lookup stretch procedure.

The procedure adopted for modifying the coordinates is illustrated in figure 3. A typical location P with coordinates (r, g) is moved along the diagonal line P_RP_Y to a position P_1 with coordinates (k, l). With P_RP_Y at 45 degrees, the amounts to be added to r and g must be equal. If the other diagonal line Q_RQ_G through P coincides with the main diagonal RG, then $r + g = 1$, $b = 0$, and P_1 must then be at P_Y. All other positions on RG are transferred to the boundary lines RY and GY, so that there is inevitably a discontinuity across the diagonal OY.

We consider the region below OY where $r > g$. The simplest way to achieve the requirement of filling the square is to set $k = r + g$ and $l = 2g$. P_1 is then at P_2, with the line Q_RP_2 parallel to the axis OG. However, it seems desirable to reduce the amount of distortion as Q_RQ_G moves closer to the origin O; a simple way to achieve this is to multiply the amount added to r and g by a factor $(r + g)$, so that P_2 moves to the revised position P_1. A complementary argument applies to points on the other side of OY, for which $r < g$. Thus, we have

$$\text{if } r > g, \quad k = r + g(r + g), \quad l = g + g(r + g); \tag{1}$$

$$\text{if } r < g, \quad k = r + r(r + g), \quad l = g + r(r + g); \tag{2}$$

Then $K = nk$ and $L = nl$.

Figure 2 illustrates the resulting colour distribution of the lookup table produced by equations (1) and (2). This square distribution doubles the number of possible (K, L) coordinates, reducing the number of combinations of R, G and B that can equal any one (K, L) coordinate. This is advantageous as it spreads the colours out more widely, allowing for better colour discrimination, especially the red, yellow and green which are important for the application to strawberry ripeness. The white point, at which $R = G = B$, is at the position $K = L = (5/9)\ n$. An effect of stretching the colour distribution into a square and rounding to obtain integer coordinates is that two blank regions exist, in the green corner where $K = 1$ and $L = 210$ to 254, and in the red corner where $K = 210$ to 254 and $L = 1$. These regions are close to where $R = B = 0$ and $G = B = 0$ respectively, so they have no effect on the functioning of the lookup table.

The advantage of using a 2D lookup table is that it allows a more comprehensive lookup than just using a one-dimensional lookup, e.g. using just the red value by itself or the CIELAB a* value [10]. It requires less computational memory to hold than a three-dimensional lookup table [14] and from a user experience perspective is simpler to use, as manually defining and visualising a 2D region on a 2D table is easier than a 3D region on a 3D table.

2.4. *Lookup table seeding and application*

Each area that needs identification (ripe, unripe, etc.) is assigned a number. The numbers used in this study are shown in Table 1. The appropriate elements of the lookup table are then seeded with the area's associated number. This seeding can be carried out either via manual selection of the appropriate elements, or by feeding a suitable seed image of a strawberry's surface through equations (1) and (2) and allocating the assigned number to the element with the calculated *KL*-coordinate.

When using the seed image method, the locations of more than one area in the lookup table may overlap. In this case the elements in the overlapping regions are seeded with the sum of the numbers assigned to the areas involved. Assigning the numbers as in Table 1, or in a similar manner, means that each area, including the overlapping ones, is uniquely identified.

Additionally, the condition that $(R+G+B) > 3$ for a pixel in the seed image was implemented to prevent nearly-black pixels, e.g. those with $R = 1$ and $G = B = 0$, from being used in the lookup table. This was done to reduce the effect of noise in the seed image triggering false identification of areas when the lookup table was applied.

Table 1. Numbers assigned to each area

Region	Ripe	Unripe	Ripe + Unripe	Calyx	Ripe + Calyx	Unripe + Calyx	Ripe + Unripe + Calyx
Assigned Number	1	2	3	4	5	6	7

To use the lookup table, equations (1) and (2) are applied to all the pixels in an image of the strawberry under test. The image then becomes segmented into different areas based on where the original pixels are calculated to lie in the lookup table. The ratio of the segmented ripe to unripe areas provides a grading for the strawberry ripeness, i.e.

$$\text{Estimated ripeness} = \text{Number of ripe pixels} / \text{Number of unripe pixels.} \quad (3)$$

In the implementation demonstrated here, if a pixel is calculated as belonging to an overlapping region in the lookup table, it is ignored. For the unripe and calyx regions whose colours are similar and where substantial overlap is expected, ignoring the overlap between them will reduce the amount of false identification of calyx and leaves as unripe, resulting in a lowering of the estimated ripeness. Refinement of the regions may be carried out for example by applying a best fit algorithm to define continuous regions for each area.

2.5. Strawberry image samples

Strawberry images were acquired at an outdoor, commercial strawberry farm using a commercially available sensor. Lighting conditions were natural daylight with slight variations in intensity due to passing clouds, and sensor's settings for white balance and exposure were left on auto. The strawberries were growing under polytunnels made of clear plastic sheeting in elevated trays organised into rows approximately 50 cm apart. To seed the lookup table, six strawberries with varying degrees of ripeness (Figure 4) were picked from a row and photographed. An image editor was used to crop the ripe, unripe and calyx areas of the photograph into separate images (Figure 5).

Subsequent photographs of strawberries growing on the plants outdoors were taken over the course of a few hours. Ten suitable strawberries from the photographs were chosen and cropped, and were used as test images for the colour lookup table method.

Figure 4. Six strawberries selected from the rows of trays.

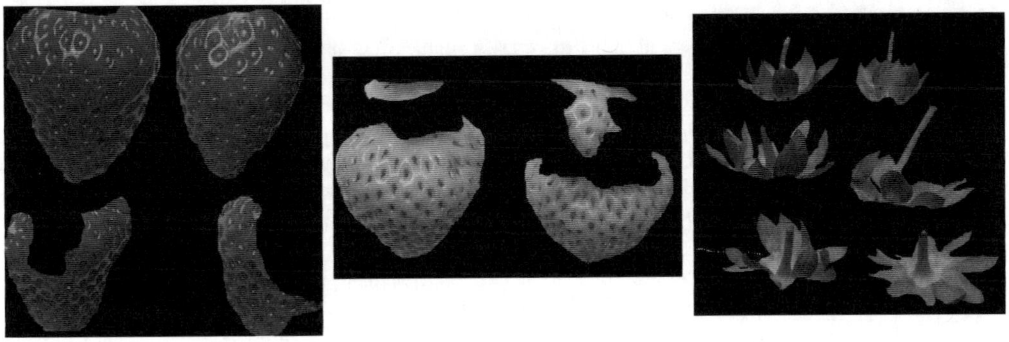

Figure 5. The cropped regions used to define the ripe, unripe and calyx regions of the lookup table.

3. Results and discussion

3.1. Daylight lookup table

The lookup table seeded with the images in figure 5, with each area colour-coded, is shown in figure 6. The three primary areas, ripe, unripe and calyx, exist in distinct locations in the table. Some overlap is seen between these three areas. The overlap between the unripe and calyx areas is larger than that between ripe and unripe due to their greater similarity in colour. Stretching the lookup table into a square has the beneficial effect of reducing the percentage area of "Ripe + Unripe" region from 7.4% of all occupied (K, L) coordinates to 4.2%, allowing for better discernment between ripe and unripe. Here we used the lookup table as it stands. Further improvements such as refining the shapes of the regions with a best-fit algorithm and better handling of overlapping regions are not implemented in this study.

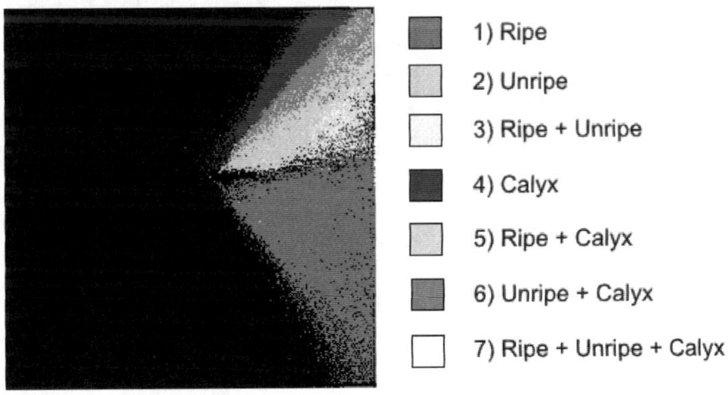

Figure 6. Lookup table created using the images from figure 5.

3.2. Image segmentation with lookup table

Figure 7 shows ten images of strawberries captured at the commercial farm with a complete range of variation in ripeness and colouration. The images contain only isolated berries to ensure only a single object is analysed per image

Examining images 1, 2 and 9 in figure 8, the lookup table successfully segments the ripe areas. In images 7 and 8 where the strawberry has both ripe and unripe regions, both areas are also clearly defined. Unripe areas are also identified correctly in most of the test images, with the exceptions of 4 and 6, where the strawberries are highlighted as being part of the overlapping unripe and calyx region of the lookup table. This region of the lookup table also has a small effect on the detection and segmentation of the calyx areas, seen in image 1. In images 8 and 9 there are calyx areas that are unsegmented due to their colours lying outside the calyx region defined in the lookup table.

Ideally images presented to the ripeness method presented in this paper will contain a single berry and this requires a good shape identification method from the computer vision system. However, in practice, some background and extraneous features will also appear in the image frame. Analysing these non-berry components in the images, it is apparent that the lookup table identifies much of the background as being unripe (green), calyx (blue) or both (purple). As the images were taken while the strawberries were still attached to the plant, this is expected because much of the background will be leaves, grass or other unripe strawberries. Grey plant supporting trusses and rope used at the test farm is segmented as unripe, due to its ratios of the (R, G, B) values. Finally, in images 1 and 3, black plastic lining is picked up as a ripe area due to the specific ratio of red to green and blue in the pixels. Some of these problems could possibly be reduced by ignoring dark pixels, during the seeding of the lookup table.

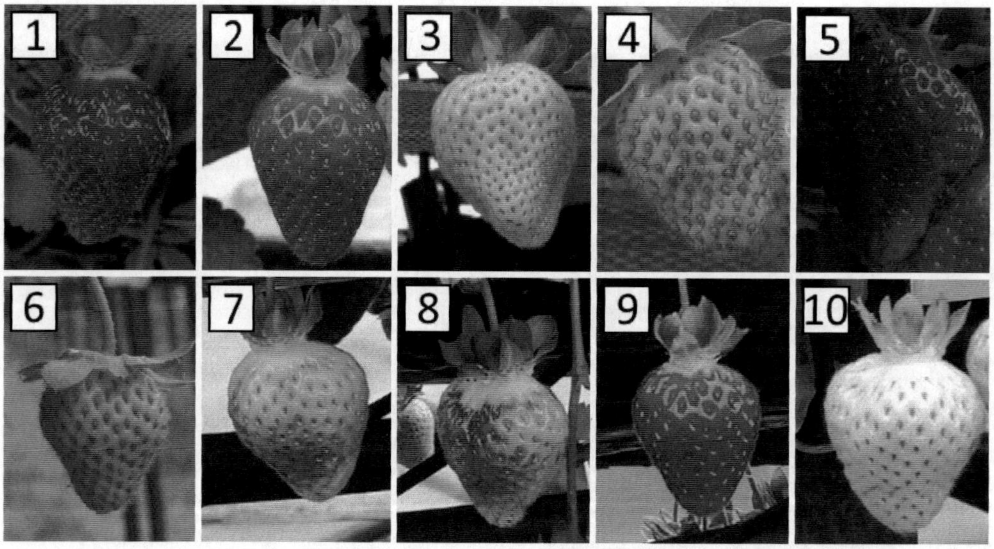

Figure 7. The cropped strawberry images used to test the lookup table method.

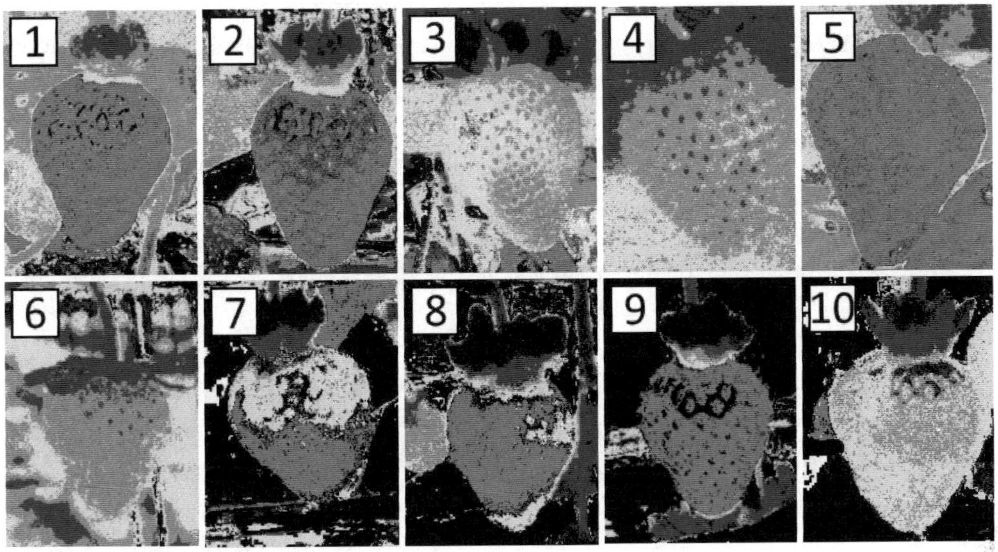

Figure 8. Images from figure 7 segmented using the same colour scheme as defined in figure 6.

3.3. Ripeness measurement

The calculated ripeness's presented in Table 2 show that the lookup table method reliably classifies the strawberries into two classes. The four ripe strawberries have values greater than 0.5 and the four unripe strawberries have values below 0.5, which compares favourably with previous works using colour cameras [10], multispectral cameras [11] and hyperspectral cameras [9, 12]. However, the accuracy of the ripeness determination is reduced in some images due to incorrect segmentation of background features, seen in the values obtained for images 1 and 2 compared to 5 and 9.

Table 2. Estimated ripeness of the strawberries shown in figure 7 and figure 8.

Strawberry	State	Estimated Ripeness
1	Ripe	0.58
2	Ripe	0.57
3	Unripe	0.11
4	Unripe	0.00
5	Ripe	0.82
6	Unripe	0.00
7	Semi-ripe	0.52
8	Semi-ripe	0.67
9	Ripe	0.84
10	Unripe	0.05

Improvements in algorithm ripeness estimation could be achieved by manual image cropping to select only the berry, which for a fully automated system would require accurate shape identification, an area which is well developed but out of scope in this work. Similarly, other ripeness detection methods have not dealt with extraneous features in the images as typically harvested strawberries were placed on a uniform colour background [9-12]. Improvements to the coordinates of lookup table can be refined through machine learning as large numbers of samples are processed in a real environment.

The regions in figure 6 are bounded approximately by straight lines radiating from a point near the central white point. This suggests that an adequate table with solid blocks of values could be constructed, based on the seeded one. Equivalently, the K and L values could be segmented numerically by evaluating $(L - L_0) / (K - K_0)$, where (K_0, L_0) are the coordinates of the origin point.

The demonstrated work uses the lookup table in a static fashion, with the regions defined from a single image taken under one set of lighting conditions. Further work would be needed to adapt the system to be used robustly under the varying daylight conditions experienced at an outdoor farm.

4. Conclusions

In this paper we have presented and demonstrated the use of a 2D colour lookup table to segment a strawberry image and use the ratio of unripe to ripe areas to estimate strawberry ripeness. Using images taken at a UK outdoor strawberry farm, the lookup table was seeded so that regions for ripe strawberry, unripe strawberry and the strawberry calyx were defined. Testing sample images of unripe, half-ripe and ripe strawberries has shown that the approach is viable. The four ripe strawberries in the sample were all calculated to have ripeness estimates greater than 0.5, while the four unripe strawberries gave estimates less than 0.5. The estimates were found to be highly affected by the presence of background features in the sample images, which were primarily misidentified as unripe regions. Suggestions on how to refine the regions of the lookup table are given as well as extensions for real-time use more appropriate for outdoor application.

Acknowledgements

This work was funded and supported by Innovate UK as part of the "AUTOPIC" project (reference number: 101814) and the UK Government's Department for Business, Energy and Industrial Strategy (BEIS) as part of the National Measurement System programme.

References

[1] Amer Eissa A H and Abdel Khalik A A 2012 Understanding Color Image Processing by Machine Vision for Biological Materials *Structure and Function of Food Engineering* ed A Eissa (London: InTech)

[2] Seema, Kumar A and Gill G S 2015 Automatic Fruit Grading and Classification System Using Computer Vision: A Review *2015 2nd Int. Conf. on Adv. in Computing and Communication Engineering* 598–603

[3] Sun D-W 2016 *Computer Vision Technology for Food Quality Evaluation 2nd ed* (Elsevier)

[4] Li P, Lee S and Hsu H-Y 2011 Review on fruit harvesting method for potential use of automatic fruit harvesting systems *Procedia Eng.* **23** 351–366

[5] House of Commons Environment, Food and Rural Affairs Committee 2017 *Feeding the nation: labour constraints* (London: House of Commons)

[6] Perkins-Veazie P 2010 Growth and Ripening of Strawberry Fruit *Hortic. Rev.* ed J Janick (John Wiley & Sons, Ltd) chapter 8 pp 267–97

[7] Pathare P B, Opara U L and Al-Said F A-J 2013 Colour Measurement and Analysis in Fresh and Processed Foods: A Review *Food Bioprocess Tech.* **6** 36–60

[8] Hunt R W G and Pointer M R 2011 *Measuring Colour 4th ed* (Wiley)

[9] ElMasry G, Wang N, ElSayed A and Ngadi M 2007 Hyperspectral imaging for non-destructive determination of some quality attributes for strawberry *J. Food Eng.* **81** 98–107

[10] Liming X and Yanchao Z 2010 Automated strawberry grading system based on image processing *Comput. Electron. Agric.* **71** S1-S66

[11] Liu C, Liu W, Lu X, Ma F, Chen W, Yang J and Zheng L 2014 Application of multispectral imaging to determine quality attributes and ripeness stage in strawberry fruit *PLoS ONE* **9** e87818

[12] Zhang C, Guo C, Liu F, Kong W, He Y and Lou B 2016 Hyperspectral imaging analysis for ripeness evaluation of strawberry with support vector machine *J. Food Eng.* **179** 11–8

[13] Hård A and Sivik L 1981 NCS – Natural color system: a Swedish standard for color notation *Color Res. Appl.* **4** 129–138

[14] Gunawardena C A, Clark L J and Dennis T J 1991 A Spot-Type Defect Detection and Colour Identification System for Agricultural Produce *1991 Proc. IECON '91* **3** 2531–34

IOP Conf. Series: Earth and Environmental Science **275** (2019) 012004 doi:10.1088/1755-1315/275/1/012004

Exploring the impact of innovation adoption in agriculture: how and where Precision Agriculture Technologies can be suitable for the Italian farm system?

G Bucci[1], D Bentivoglio[1]*, A Finco[1] and M Belletti[1]

[1] Department of Agricultural, Food and Environmental Sciences (D3A), Polytechnic University of Marche Region, Brecce Bianche 60131, Ancona, Italy
*email: d.bentivoglio@univpm.it

Abstract. Starting from the twenty-first century, the digital paradigm has brought with it a new way of thinking about innovation within the firms through the introduction of the Internet of Things (IoT) and the Information and Communications Technology (ICT), keys factors of digital transformation technologies. In addition, the use of Precision Agriculture (PA) has created a new approach to farming practices, allowing farmers to reduce the inputs, while protecting natural resources. Today, also in Italy PA is starting to be an effective technological innovation potentially capable to improve farms efficiency and sustainability. Given these considerations, the present paper aims, firstly, to review the factors more affect the adoption of precision agriculture technologies (PATs); secondly, to investigate where the PATs could be convenient to apply in Italy. A multi-Criteria Decision Analysis method is performed to achieve this goal. The analysis shows that the geographical area more likely to use PATs in Italy is the North East territory. However, in Italy, the adoption of these technologies in agriculture meets various obstacles: the cultural barrier to innovation and a limited awareness of benefits; the reduced average size of farms, with the difficulty of investing in and appreciating the benefits of PATs.

1. Introduction

In Europe innovation is a hot topic and for this reason it has been placed at the heart of the EU's policies. In the interest of monitor the innovativeness of Members State, EU created a set of indices based on economic, demographic and social performance. According to European Innovation Scoreboard (ESI) 2018 [1], at worldwide level, the most innovative countries are South Korea, Canada and Australia, followed by Japan and United States (Figure 1).

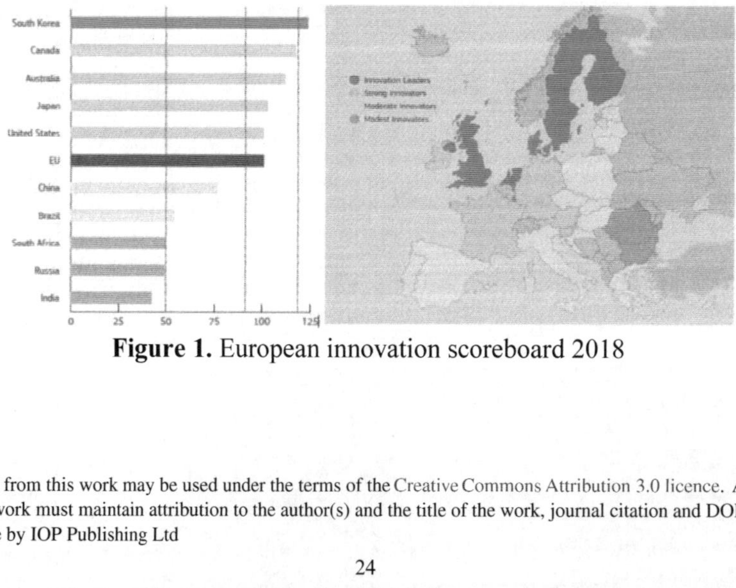

Figure 1. European innovation scoreboard 2018

Europe has lasted at 6th position, China lies in the 7th position, Brazil in 8th. At European level, Sweden remains the EU innovation leader, then Lithuania, the Netherlands, Malta, the UK, Latvia, and France are the fastest growing innovators. Finally, Italy still remains a moderate innovator. Within the agricultural scenario, in the last 100 years, innovation through technology has been one of the major factor changing agricultural sector [2, 3, 4]. The concept of innovation is a very broad concept and it has been studied extensively [5, 6, 7]. Thus, during the years, the literature on innovation has coined its own vocabulary and various classification. Focusing on agriculture and according to Sunding and Zilberman (2001) [8], the classification of innovation can be distinguished in three categories, based on:

- policy questions and the forces behind the generation and the adoption of innovation;
- form distinguished between process innovation and product innovation;
- their impact on economic agents and market.

Innovations can fall into one classification or fall into several of these categories. In addition, according to Rogers' Theory (1962) [9], the adoption of an innovation follows an S curve when plotted over a length of time. The categories of adopters are: innovators, early adopters, early majority, late majority and laggards (Figure 2).

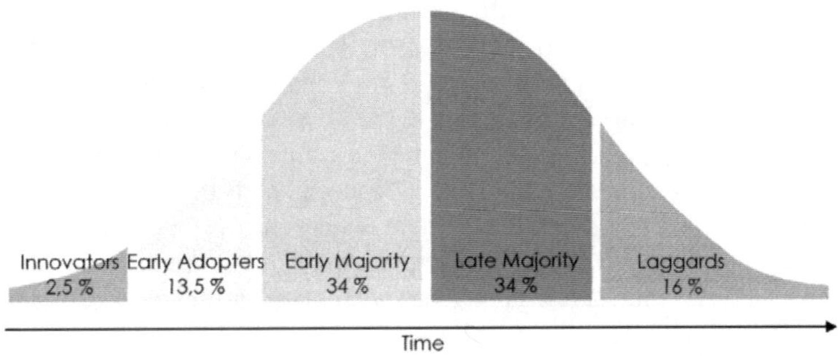

Figure 2. The model of diffusion of innovation by Rogers (1962)

According to Rogers, the causes of poor agricultural performance were essentially due to a low technological development and could be solved by developing and improving these lacks [10]. However, in general, innovation is a complex process and the adoption of new technologies in agriculture is not often immediate [11]. Indeed, the difficulty to apply innovative technologies is due to because farming operators go to meet complex decision-making problems. In order to optimize agricultural process/operations, farmers need to consider a variety of conditions and inputs (water, nutrients, fertilizers, weather, etc.) that can influence the entire agricultural system. Finally, the institutional setup is another important factor to be understanding to innovation activity. In particular, agricultural policy that supports the income of farmers could influence the innovation process. In fact, there is a strong connection between public support for innovation research and programs that support farm income. This could explain the investment in innovation in the agriculture sector.

In this context, thinking about the need of new innovative technologies in agriculture, precision agriculture (PA), is recognized as a major contribution to farm sustainability and innovativeness [12]. PA is a relatively new concept and many definitions of this agriculture exist in literature. In addition, many people have different ideas of what PA should encompass. PA is a "management strategy that uses information technologies to bring data from multiple sources to bear on decisions associated with crop production, designed to increase long term, efficiency, productivity and profitability while minimizing unintended impacts on wildlife and the environment" [13]. The goal of PA research is to "define a decision support system (DSS) for whole farm management with the goal of optimizing returns on inputs while preserving resources" [14]. It is a site-specific management which allows to do the right practice at "the right location and time at the right intensity" [15]. As Auernhammer and Demmel (2015) [16] highlighted, PA can be considered one of the different types of precision land management, which also includes precision forestry and aquaculture. Since 1980s, PA has developed from the use of the first

GPS system and yield maps to the latest applications with controlled traffic farming with RTK systems and UAVs (Unmanned Aerial Vehicles). Nowadays, more advanced technological systems are being developed, such as drones, agricultural robots and Decision Support Systems, so that precision agriculture technologies (PATs) became almost commonplace in America and Australia, compared to Europe and Italy, in which their application is still limited. Nevertheless, the percentage of PA adoption has also increased in Europe, with a rate of 15-20%, and today, also in Italy PA is starting to be an effective technological innovation potentially capable to improve efficiency and sustainability. Given these considerations, the present paper aims to collect the variables that affecting PATs adoption in agriculture and also to investigate where the PATs could be convenient to apply in Italy.

2. Data and Methods

The need for a smarter, more technological agriculture is rapidly spreading among farmers, however several studies came to the conclusion that the adoption of PA is limited both in the USA and in Europe [17,18,19,20,21,22]. Based on the available literature, several factors are found to be significant in affecting the adoption of technological innovations in the agriculture system. These factors are mainly related to the characteristics of the farms and the characteristics of the farm owner. The first factor to consider in determining farmers' propensity to adopt PATs is the computerization of farms, understood as the computerized management of business processes, which in Italy concerns only 4% of farms. Land size is the second factor to be taken into account: a large land size is usually associated with a higher propensity to adopt Smart Farming Technologies [23, 24]. In fact, larger farms have a higher need for efficiency and this usually induces them to adopt PATs but they also have the possibility to better amortize technology investment costs [25]. This is especially true in USA where the average farm occupies an area of 180 hectares, compared to Europe, where the average stands only on 16 hectares. In parallel, even the presence of Agricultural contractors is relevant in the process of PATs adoption: this is because agricultural contractors have a general knowledge of farming methods and the types of farm machinery used in these operations and generally are used to work in large size farms, so they represent an interesting market segment to catch by technology providers [26]. In fact, contractor services will be used to share new technologies in agriculture, especially in Italy in which are contractors more widespread and in 2017 they arrived to manage the 1/3 of the total Italian farms and the 40% of the Utilised Agricultural Area. Finally, considering the farmer owner's characteristics, in accordance with Rogers' Theory, age and education are the most influential factors in the innovation process: younger farmers, the Innovators, show a more innovative attitude, compared to the oldest, the Laggards [27]. In Italy the rate of aging in agriculture is very high compared to other countries: farmers over 65 years represent 40% of the total, in France over 65 are 12%, while in Germany they are only 5%. This demographic structure is obviously reflected on the generational turnover index (the relationship between entrepreneurs aged under 35 and entrepreneurs over 65), which sees Italy in a position of relative weakness compared to the main European countries. Similarly, also the Educational level may influence the probability of investing in PA, because of a higher education may be helpful in enhancing the perceived benefit and usefulness of Smart technologies. In fact, the farmer's education positively influences even the number of precision technologies adopted. In particular, as Paxton et al., (2010) [28] highlighted, just one additional year of schooling, may correspond to an 8 percent increase in the number of precision technologies adopted. A higher educational level may correspond with higher abilities to utilize these new technologies, with higher willingness to pay for them. Farmers with a higher level of education will be abler to correctly interpret information from Big Data and to more understand the same technology language. However, one of the major limitations in the Agrifood sector is the lack of information and therefore, the low awareness of the availability of new farm innovation and of their potential benefits. In this sense, education could provide the right incentive to challenge agriculture practice, by inducing farmers to adopt PATs and raising their willingness to pay for these technologies. It is important to highlight that there are other important factors that could influence the adoption of innovation technology. A significant role is given by the policy that can support the innovation adoption through different measures and programs. In addition, the diffusion of innovation technology could be influenced by aggregate adoption. The aggregation is seen as one way to increase the chances to compete by generating synergies and play an important role in innovation processes [29]. Precision agriculture management is thus by definition a multi-objective decision-making process that must take into account of different criteria, opinion, data, preference and objective [30, 31]. Considering this definition, it is important to assume that a considerable amount of aspects will be required to understand the feasibility and the possibility to apply this new technology in the agricultural sector. To date, Multi-Criteria Decision Analysis (MCDA) methods are incorporated between the examination and evaluation techniques considered useful by the EU and it is seen as a methodology to identify and support choices aimed at obtaining a sustainable development process [32, 33, 34, 35]. In general, the MCDA is organized in two steps [36]. The first phase includes the construction and compilation of the evaluation matrix, according to the established objective/problem. This matrix consists of the different alternatives, various criteria and different weightings. The second phase regards the processing of the data in the evaluation matrix used to assess the alternatives via a variety of different procedures. In details, the MCDA method described in this study has been used, in conjunction with factors (Criteria) collected by literature relating to the main variables influencing the adoptions of PATs, for identification of geographical areas (alternatives) more likely to use these technologies in Italy.

The MCDA in our study, was carried out using the software 'Definite 3.1' [37]. In details, the aim was to identify among the different geographical area of Italy (North East, North West, Central, South and Islands) the best alternative in which apply innovation technologies. The alternatives considered were identified on the basis of the different precursory criteria for the adoption of PATs that are: the farms size (>20 ha), the use of contractors, the computerization of the farm (digitalization/digital farms), the age (between 18-40) and the degree of education (qualifications). All the data were obtained from ISTAT census 2010. Therefore, a summary table (Table 1) was created to represent our evaluation matrix, which shows the average values for each criterion chosen for the different alternatives. The assignment of weights given to the relevant criteria in the multi-criteria analysis depends on the type of evaluation to be performed.

Table 1. Evaluation matrix

ALTERNATIVE	CRITERIA				
	Farms (>20 ha) %	Contractors %	Digital farms %	Age 18-40 %	Qualifications %
North East	18	2	9	13	26
North West	14	3	11	11	20
Central	13	1	4	8	28
South	7	1	1	9	24
Islands	18	1	2	13	23

3. Result and Discussion

After the design and quantification of the evaluation matrix, the data entered in the matrix were standardized per column. The method used for the all criteria is defined as 'maximum standardization'. In this context, an equal weight was given to the farms size (>20 ha), the use of contractors, the age (between 18-40) and the degree of education (Qualifications) because they were considered of equal importance. On the contrary, a different weight is given to the computerization (Digital farms) because this factor has been identified by the literature as the main precursor factor for the adoption of PATs. According the studies of Daberkow and McBride (2003) and Isgin et al. (2008) [38, 39], the use of the computer use is a predictor for the adoption of PATs, because it is supposed that a farmer who already uses a computer to manage company data will be more inclined to adopt a more developed technology in the near future. The results from our MCDA are shown in a graphic form (Figure 3). The analysis shows that the geographical area more likely to use PATs in Italy is the North East territory.

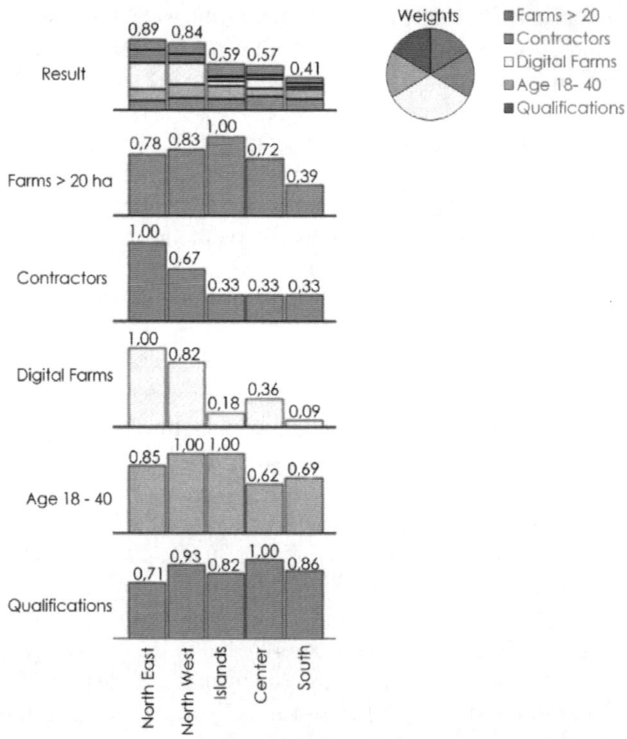

Figure 3. Multi-criteria analysis elaboration

This area, in fact, recorded the best performance for two criteria including computerization that assumes a much higher value than the other criteria. However, also the North West register a good performance, followed by Islands and central territory that show similar performance. These results are confirmed by the various applications of PA in Northern Italy [40, 41], which mainly concern rice cultivation and fruit trees. Relating to the Islands, these territories have good potential to apply these new technologies. Indeed, Islands recorded the best performance for two criteria that are big land size and younger farmers. Nevertheless, the presence of digital farms is low, even if different projects with PATs are already being applied [42]. For these reasons, the digital providers could decide to invest in these geographical areas. Finally, the South is positioned at the last place because of it register the lowest level of computerization.

4. Conclusion

Today, also in Italy PA is starting to be an effective technological innovation potentially capable to improve efficiency and sustainability of agriculture. However, Italy continues to suffer a considerable delay in the use of PATs compared to most European Union countries (Currently, only 1% of Italian utilized agricultural areas are managed through PA technologies). Despite the lack of propensity for the use of technologies in farms, the times seem ripe for a technological change in the Italian agricultural context; and this is evidenced by the fact that, according to the Smart Agrifood Observatory, among the 480 new firms supplying technologies for agriculture born since 2011 to date worldwide, 15% are Italian firms providing technological solutions to improve efficiency and sustainability for agricultural production. It is clear that on the market the increasing offer of technology for agriculture does not match the demand for ICT from farmers. Italy is at the top for available technology, but the farms are, not "smart". This study highlights that the adoption of PATs depends on different factors, especially the computerization. In addition, the limited average size of farms, with the difficulty of investing in and appreciating the benefits of precision technologies represent a barrier to innovation. Finally, most of the farmers already using this technology face many difficulties in managing such a large amount of data collected by PATs and in using technologies efficiently. The main problem is how to decode the data since there exist many interpretations. In the next years, it will be meaningful to develop more simple practical technologies in order to support the farmer in making decisions and educate farms to innovate. Of course, this would not be possible without the support of policies that enable innovations to materialize.

References

[1] Hollanders H and Es-Sadki N 2018. *European Innovation Scoreboard 2018* Publications Office of the European Union, Luxembourg
[2] Schultz T W 1964 *Economics* **37** (1): 47-61.
[3] Cochrane W W 1979 *The development of American agriculture: A historical analysis* U of Minnesota Press.
[4] Reardon T, Echeverría R, Berdegué J, Minten B, Liverpool-Tasie S, Tschirley D and Zilberman, D 2018 *Agricultural Systems*. (In Press)
[5] Touzard J M, Temple L, Faure G and Triomphe B 2015 *Journal of Innovation Economics Management* **2**: 117-142.
[6] Marotta G, Nazzaro C and Stanco M 2017 *Journal of Globalisation and Small Business* **9** (2-3): 144-167.
[7] Blanc S, Accastello C, Girgenti V, Brun F and Mosso A 2018 *Calitatea* **19** (165): 139-142.
[8] Sunding D and Zilberman D 2001 *Handbook of agricultural economics* **1**: 207-261.
[9] Rogers E M 1962 Diffusion of innovations (1st ed.) New York Free Press.
[10] Guerin L J and Guerin T F 1994 *Australian Journal of Experimental Agriculture* **34** (4): 549-571.
[11] Pierpaoli E, Carli G, Pignatti E and Canavari M 2013 *Procedia Technology* **8**: 61-69.
[12] Aubert B A, Schroeder A and Grimaudo J 2012 *Decision Support Systems* **54**: 510–520.
[13] National Research Council 1997 *Precision Agriculture in the 21st Century: Geospatial and Information Technologies in Crop Management.* Washington, D.C., USA: National Academy Press.
[14] McBratney A, Whelan B, Ancev T and Bouma J 2005 *Precision agriculture* **6** (1): 7-23.
[15] Gebbers R and Adamchuk V I 2010 *Science* **327** (5967): 828-831.
[16] Auernhammer H and Demmel M 2015 *State of the Art and Future Requirements.* In Qin Zhang (Ed) Precision Agriculture Technology for Crop Farming CRC Press USA.
[17] Daberkow S G and McBride W D 2003 *Precision agriculture* **4** (2): 163-177.
[18] Fountas S, Blackmore S, Ess D, Hawkins S, Blumhoff G, Lowenberg-Deboer J and Sorensen C G 2005 *Precision Agriculture* **6** (2): 121-141.
[19] Pedersen S M and Lind K M 2017 *Precision Agriculture: Technology and Economic Perspectives* Springer International Publishing.
[20] Bucci G, Bentivoglio D and Finco A 2018 *Calitatea*, **19** (S1): 114-121.
[21] Barnes A, De Soto I, Eory V, Beck B, Balafoutis A, Sánchez B and Gómez-Barbero M 2019 *Environmental Science & Policy* **93**: 66-74.
[22] Barnes A P, Soto I, Eory V, Beck B, Balafoutis A, Sánchez B and Gómez-Barbero M 2019 *Land Use Policy* **80**: 163-174.
[23] Walton J C, Lambert D M, Roberts R K, Larson J A, English B, Larkin S L and Reeves J M 2008 *Journal of Agricultural and Resource Economics* **33**(3): 428-448.
[24] Lambert D M, Paudel K P and Larson J A 2015 *Journal of agricultural and resource economics* **40** (2): 325-345.
[25] Castle M H, Lubben B D and Luck J D 2016 *Factors influencing the adoption of precision agriculture technologies by Nebraska producers*. University of Nebraska-Lincoln Digital Commons Presentations Working Papers and Gray Literature Agricultural Economics. 49
[26] Taylor J A, McBratney A B, Viscarra Rossel R A, Minasny B, Taylor H J, Whelan B M and Short

M 2006 Development of a multi-sensor platform for proximal soil sensing. In *18th World Congress of Soil Science July* (pp. 9-15).

[27] Botsiou M G, Koutsou S and Dagdilelis V 2014. *Scientific Bulletin–Economic Sciences* **13**: 2-12.

[28] Paxton K W, Mishra A K, Chintawar S, Roberts R K, Larson J A, English B C and Martin S W 2011 *Agricultural and Resource Economics Review*, **40** (1):133-144.

[29] Bentivoglio D, Giampietri E and Finco A 2016. *Quality-Access to Success* **17** (1): 57-63.

[30] Jones D and Barnes E M 2000 *Agricultural Systems* **65** (3): 137-158.

[31] Finco A, Bentivoglio D and Bucci G 2018 *Economia Agro-Alimentare/Food Economy* **20** (2):181-192.

[32] Nijkamp P and van Delft A 1977 *Multi-criteria analysis and regional decision-making* Springer Science & Business Media Volume **8**.

[33] Rehman T and Romero C 1993 *Agricultural systems* **41** (3): 239-255.

[34] Kiker G A, Bridges T S, Varghese A and Seager T P, Linkov I 2005 *Integrated environmental assessment and management* **1** (2): 95-108 .

[35] Finco A, Bentivoglio D and Nijkamp P 2012 *International Journal of Foresight and Innovation Policy* **8** (2/3): 173-188.

[36] Janssen R 1992 *Multiobjective Decision Support for Environmental Management, Environment and Management* Kluwer Academic Publishers Dordrecht Netherlands.

[37] Daberkow S G and McBride W D 2003 *Precision Agriculture* **4** (2): 163-177.

[38] Isgin T, Bilgic A, Forster D L and Batte M 2008 *Computers and Electronics in Agriculture* **62**: 231-242.

[39] Stroppiana D, Migliazzi M, Chiarabini V, Crema A, Musanti M, Franchino C and Villa P 2015 Rice yield estimation using multispectral data from UAV: A preliminary experiment in northern Italy. *In 2015 IEEE International Geoscience and Remote Sensing Symposium (IGARSS)* (pp. 4664-4667). IEEE.

[40] Nutini F, Confalonieri R, Crema A, Movedi E, Paleari L, Stavrakoudis D and Boschetti M 2018 *Computers and Electronics in Agriculture* **154**: 80-92.

[41] Zambon I, Delfanti L, Marucci A, Bedini R, Bessone W, Cecchini M and Monarca D 2017 *Agriculture* **7** (7): 56.

Design a Web Platform to manage environmental monitoring information to be used in multicriteria evaluations of Green Infrastructures

R Gallo[1], G Ristorto[2], A Bojeri[2], N Zorzi[3], M F Rinaldi[1], G Sauli[4], and F Mazzetto[1]

[1] Faculty of Science and Technology, Free University of Bozen (FUB), Piazza Università 5, 39100 Bolzano (Italy)
[2] Mavtech srl, Bolzano, 39100, Italy
[3] Maccaferri Innovation Center srl, Bolzano, 39100, Italy
[4] Naturstudio srl, Bolzano, 39100, Italy

e-mail: raimondo.gallo@unibz.it

Abstract. The aim of the WEQUAL project "WEb service centre for QUALity multidimensional design and tele-operated monitoring of Green Infrastructures") is the development of a system able to support a quick environmental monitoring of watercourses where new hydraulic structures are intended to be built, encouraging the identification of design solutions supporting the diffusion of Green Infrastructures (GI). The WEQUAL´s idea is to organize a service centre where users (designer, technician, researcher) can access the service through a personal account and carry out an assessment of alternative GI projects.
Through some automatic procedures, accessible via a suitable web platform, a set of algorithms may be applied to process raw data collected by an UAV (Unmanned Aircraft Vehicle) equipped with a 3D Lidar, multispectral camera and RGB camera, for the purpose of calculating the WEQUI Index. The WEQUI index is used to assesses the eco-morphological status of the monitored watercourse, taking advantage from data related to NDVI index, Digital Terrain Model (DTM), Digital Surface Model (DSM) and a 3D point cloud classification.
The computed value of the WEQUI index may be used to assess the eco-morphological quality at ex-ante and ex-post river stabilization or protection intervention. The platform will provide a shared environment integrating indices calculations and environmental-parameters assessment for multidimensional evaluations.

Keywords: Information Management, Field Surveys, Data Logging, Multidimensional Evaluations, River Ecosystems, Environmental Monitoring

1. Introduction

In the last period, European Union (EU) increased its interest in Green Infrastructures (GI), especially in river environments, in order to safeguard biodiversity of ecosystems, with the principal aim to ensure proper safety levels to people, villages and infrastructures. For this scope, the European Environmental Agency (EEA) has stated different policies and directives on this topic, such as the Water Framework Directive (WFD-2000/60/EC) and Flood Directive (2007/60/EC) [1, 2]. The Water Framework

Directive aims at classifying water bodies and to identify anthropogenic impacts on them, in order to carry out qualitative and quantitative evaluation of the improvement of river systems, while the Floods Directive focuses on the design and planning phases of hydraulic works, with the aim of reducing the risk of natural disasters due to floods, landslides or erosion and safeguarding aquatic ecosystems [3]. In this way, EEA works towards ensuring high standards of ecological quality, biodiversity and riparian functionality in all riparian lands. The respect of those principles has become a key issue for technicians and specialists involved in river management.

The aim of the WEQUAL project ("WEb service centre for a QUALity multidimensional design and tele-operated monitoring of GI"), developed withing the European EFRE-FESR Südtirol-Alto Adige framework, is the development of a new decision-support system, able to support a quick environmental monitoring of watercourses where new hydraulic structures are intended to be built. The system includes a data analysis service able to provide promptly the ecological indices usually required by the EU directives, which can be easily used then within multidimensional evaluation approaches, aiding decision-makers featured by different standpoints. In short, the project´s idea is to organize a service center managing both the web platform and the whole data collection and analysis process, in order to quickly and automatically assess the environmental impact of lateral and transverse hydraulic structures. Thus, the final user of the product, such as technicians, administrators or researchers, can use an innovative tool as a decision-support system. Through a personal account, the user can access the service and require the evaluation of alternatives GI projects, minimising in-field activity. Then, all the required assessments will be sent to the user via web by the service center.

2. Material and Methods

The paper describes the approach followed to collect and process data acquired by automatic survey carried out through remote sensing applications. For this task, a UAV equipped with different sensors has been specifically developed and used for raw data collection. All those data are then post-processed by interpretative algorithms, to obtain information for the evaluation of the required eco-morphological indices. Al the implemented algorithms run in MATLAB or Phyton environment.

2.1. The methodology

Within the WEQUAL project, a new assessment method was developed, aimed at evaluating the eco-morphological quality of watercourses. The method is called WEQUI (WEQUAL Eco-morphological QUality Index) and is based on the analysis of the 15 quality indicators listed in Table 1. The method is multidisciplinary and applicable to both highly-natural watercourses and artificialised watercourses (i.e. characterised by the presence of artificial elements, as for example hydraulic structures or infrastructures, within the river corridor).

The WEQUI method was inspired by some existing methods described in the literature, which were used as a reference to recognise those elements that are the most relevant for eco-morphological characterisation of watercourses [3, 4, 5, 6, 7, 8]. In this way, 13 quality indicators were identified, concerning multiple aspects related to bio-physical aspects. Then, further two indicators were added to complete the method. These two indicators include a kind of assessment of the carbon emission and absorption observed during the whole life cycle of artificial structures, i.e. from the production/construction phase to the full-operating condition, when a new equilibrium is reached by the local ecosystem after construction.

WEQUI is intended to be used for two main purposes: (a) evaluating and monitoring the condition of homogeneous portions of a watercourse, leading to the identification of positive and critical elements affecting eco-morphological quality, and (b) estimating long-term eco-morphological alterations introduced by artificial elements, either present or designed.

In both cases, indicators are intended to be assigned according to both qualitative and quantitative information. Profitable quantitative information can be obtained through remote-sensing surveys, using proper sensors, as described in the next Sections. High-resolution data collected in this way can be processed profitably, applying some custom algorithms specifically implemented within the WEQUAL project (see Section 3) and able to provide values of some significant variables and indices (e.g.: cross-section elevation and slope, vegetation metrics, NDVI). Such variables and indices can be used to guide effectively the attribution of several indicators of the WEQUI method. The assessment procedure leading to the calculation of the WEQUI index is integrated in a tailored web platform, conceived according to the Software-as-a-Service approach and currently under development. The web platform will be accessible through a simple login (registration will be free of charge). Such a tool is intended as a support for users applying the WEQUI method for both the (a) and (b) purposes.

Table 1. Indicators to be evaluated for the eco-morphological assessment of fluvial areas. They are divided into two categories: indicators which assessment can be based on remote-sensing data processing and indicators that require necessarily a different type of assessment.

Indicators assessable through remote-sensing data processing	Indicators differently assessable
1) Land use	3) Vertical continuity
2) Lateral continuity	7) Hydrologic regime
4) Longitudinal continuity	8) Chemical quality
5) Morphological heterogeneity	9) Macrobenthos community
6) Retention capability	15) Carbon footprint
10) Fish suitability	
11) Riparian strip vegetation	
12) Riparian strip width	
13) Riparian strip continuity	
14) Carbon sequestration	

2.2. Raw data acquisition

Within the project, different Unmanned Aerial Vehicles (UAVs) have been developed to collect data within riparian areas. UAVs are increasingly used in environmental monitoring due to their highly flexibility [9]. The flying platforms developed for WEQUAL show different configurations (fixed-wing and rotary-wing) and, therefore, different flight performances.

The AGRI-1900 (Figure 1a) is a fixed-wing UAV. This kind of configuration has a very high aerodynamic efficiency, resulting in higher flight endurance (typically 30 ÷ 45 min). Since the survey speed is quite high (12 ÷ 15 m/s – 43 ÷ 54 km/h), this platform is able to cover large area (25 ÷ 50 ha) with a single flight. The autopilot onboard increases the aircraft stability and aids the pilot in controlling the vehicle. The main drawback of the fixed wing configuration is that a wide area free from obstacles is necessary for take-off and landing manoeuvres. The flight altitude is high (100 ÷ 150 m AGL), thus the resolution of the survey is low. A fixed-wing platform returns outputs with the maximum representativeness, but with low details.

Rotary-wing UAVs have the capability to take-off and land in narrow spots as well as they are able to fly in hovering. Rotary-Wing UAVs carry out heavier payload in comparison with a fixed-wing UAV with the same Maximum Take-Off Mass (MTOM). The flight altitude is low (30 ÷ 70 m), thus the resolution of the collected data is high. Therefore, rotary-wing UAVs return the maximum detail level, but with low representativeness of the overall scenario. The main drawback of this kind of platform is the low endurance (typically 15 ÷ 20 min) and low survey speed, thus they can monitor limited area (2 ÷ 5 ha) with a single flight. The Q4E (Figure 1b) is a quadcopter and it has been specifically designed and developed within WEQUAL project, to carry an RGB camera or a multispectral camera. The DJI S900 (Figure 1c) is a multirotor with 6 rotors and it has been modified to embark a LiDAR sensor. Finally, the DJI Spark (Figure 1d) has been modified to reduce its weight below 300 g. Thanks to its low weight, it can be operated in urban areas or close to critical infrastructures (railways or highways) without a specific permit to fly (ENAC, 2018).

Table 2 shows the main features of these four UAVs and the sensors they are able to carry onboard.

Figure 1. The UAVs developed during the WEQUAL project. Each one is used for different tasks and it is equipped with specific sensors: a) MAVTech AGRI-1900 for multicamera or RGB survey for large surfaces; b) MAVTech Q4E for multicamera or RGB survey for small surfaces; c) DJI S900 for LiDAR surveys; d) DJI SPARK for video recording.

Table 2. UAVs main features and sensors

	Mavtech AGRI-1900	**Mavtech Q4E**	**DJI S900**	**DJI Spark**
Configuration	Fixed-Wing	Rotary-Wing	Rotary-Wing	Rotary-Wing
Nr. Of rotors	1	4	6	4
MTOM [kg]	2.8	3.5	8.2	0.3
Diagonal wheelbase [m]	-	0.590	0.900	0.17
Wingspan [m]	1.90	-	-	-
Endurance [min]	45	20÷25	15	15
Survey speed [m/s]	12	6	3	2÷5
Surface per mission [ha]	40	4	5	-
Sensors	RGB camera Multispectral sensor	RGB camera Multispectral sensor	LiDAR	Integrated videocamera

The Sony RX-100 is the RGB camera used in the project. It has a resolution of 20.1 MP and a 3x optical zoom. During photogrammetric surveys, the focal length is fixed to the lower value (8.8 mm), thus the angle of view and the footprint are the maximum for this camera. The multispectral sensor is the Micasense RedEdge-M. It is able to acquire, at the same time, five different images in the bands of BLUE, GREEN, RED, REDEDGE, NIR. The resolution of each image is 1 MP and, thanks to the GPS connected to the camera, the position of each image is stored in the Exif metadata. This information is very useful during the mosaicking procedure. RedEdge-M employs narrow filter that increase the sensitivity in comparison with modified camera, to capture the most relevant segments of the spectral curve and to obtain higher quality multispectral images.

Depending on the flight altitude, the RGB camera and the multispectral sensor record images with different Ground Sampling Distance (GSD). Table 3 shows the difference between the two sensors.

Table 3. Ground Sampling Distance (GSD) at different flight altitude for the RGB camera and multispectral sensor

Flight Altitude	Sony RX-100 GSD [cm/pixel]	Micasense RedEdge-M GSD [cm/pixel]
50	1.4	3.4
70	1.9	4.8
120	3.3	8.2
150	4.1	10.2

The LiDAR sensor is used to obtain the three-dimensional model of the riparian area. It exploits the Light Detection and Ranging technique to measure the distance between an object or a surface with a laser pulse. A GPS-INS platform determines the position and the attitude of the sensor. A PPK (Post Processing Kinematic) correction is performed to improve the geo-referencing of the dataset.

The LiDAR sensor employed in WEQUAL project is the Yellowscan Surveyor. It is capable to records two echoes: the first one is related to the top of the canopy, while the second one represents the soil. Depending on the flight altitude and the survey speed, the point cloud density changes, according to Figure 2. For the project purposes, 3D models with point cloud density of at least 100 points per square meters have been obtained.

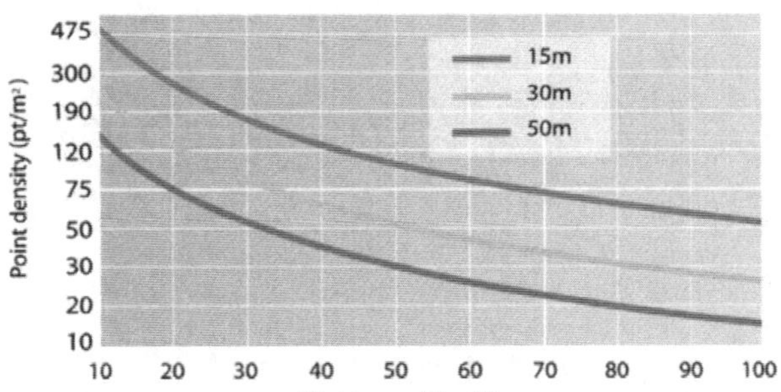

Figure 2. Point Cloud Density and survey speed for the Yellowscan LiDAR [10].

3. Results

Once remote-sensing data have been acquired and sent to the processing center, a set of ad-hoc algorithms are used to evaluate eco-morphological indices. Up to now, the implemented algorithms allow to assess just a couple of indicators among those listed in Table 1, i.e.: the land use assessment and the riparian strip identification.

The process includes the use of NDVI, Digital Terrain Model (DTM), Digital Surface Model (DSM) and a 3D point cloud classification, achieved by a LiDAR survey.

The NDVI extraction permits to distinguish among areas covered by riparian vegetation rather than by water or soil (Figure 3). At this step, only multispectral data are considered to calculate the water, vegetation and soil coverage.

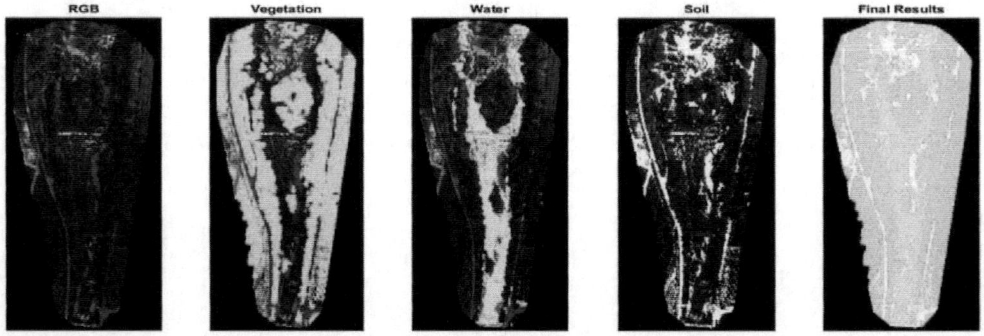

Figure 3. Graphical results obtained by the algorithm's elaboration for the distinction of the land cover. In Green, Cyan and Yellow are identified the vegetation, water bodies and soil, respectively.

Information provided by the vegetational index analysis is then processed together with data contained in the DTM/DSM maps. The digital models of the terrain are considered here to evaluate vegetation height, in order to provide a trees classification based on their height. In this work, four classes of plants are defined:

- "Grass", height from 0 [m] to 0.5[m];
- "Bushes", height from 0.5[m] to 1 [m];
- "Small Trees", height from 1 [m] to 3[m];
- "High Trees", height from 3[m] to 40[m].

At the end, the integration between multispectral data and LiDAR data generates the classification map of the investigated area differentiating water, soil and the 4 classes of vegetation (Figure 4).

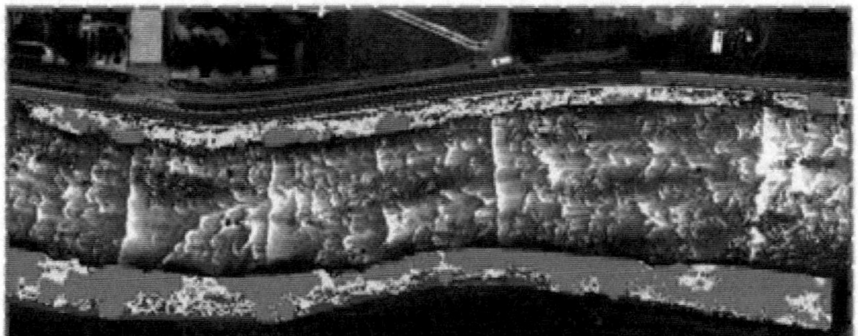

Figure 4. Graphical result of algorithm procedures for riverbanks, riparian and alluvial riparian vegetation (the green, blue and red colors refer to the herbaceous layer, riparian shrubs and riparian arboreal, respectively).

Acquired data could be considered also to support the assessment of other WEQUI indices, for example computing river sinuosity or the variability of banks and slopes or identifying the presence of artificial structures (for example transverse and lateral hydraulic structures). Processing data collected by the LiDAR survey, the algorithms can recognize river banks through the analysis of the frequency distribution of elevation and slope in the areas near the watercourse. Thus, the process identifies the riversides as well as their slope considering the cross-section of the river (Figure 5).

At the end, to provide the user a final evaluation of the eco-morphological state and impact of the riverside environment, the platform's multidimensional analysis tool combines results obtained from remote-sensing data processing and some other information in order to generate a concluding document where all the 15 eco-morphological indices of WEQUI are quantified and the quality level of the watercourse is classified.

Results produced by the automatic assessment procedures are currently under validation.

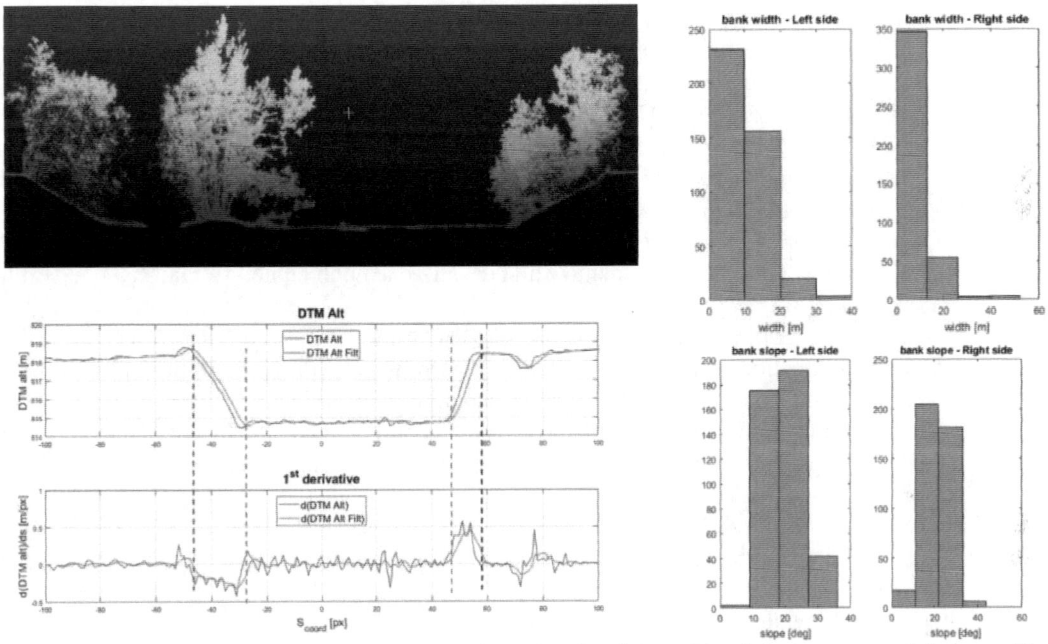

Figure 5. Exemplificative results obtained by the algorithms. Top left: a tranversal river profile obtained by LiDAR (colors represent diffent classes of heigth). Bottom left: riverbank extraction from DTM. The algorithm identifies riverbank foot and head by means of first derivative analysis of the transverse profile. The slope values of the banks are used for the physical characterization of the banks. Right: side-features of the analysed sections

4. Conclusions

In conclusion, up to now, the developed algorithms led to the computation of some useful indices, supporting the calculation of the WEQUI index within a tailored web platform. These specific indexes could be used also to carry out multicriteria analyses, even enabling the possibility of integrating additional non-technical or non-ecological indicators such as investment required to build a new hydraulic structure, maintenance costs or social acceptance of citizens. Finally, the web platform is intended to get interests of GI designers and policy makers, providing a shared environment able to integrate detection and assessment of some complex indices and multidimensional evaluations supported by an expert guide.

Acknowledgements
Research funded by the program EFRE-FESR 2014-2020 for South Tyrol – WEQUAL Project – CUP I52F16000840005.

References

[1] EU. Directive 2000/60/CE of 23/10/2000. Water Framework Directive. EC OJ n. L 327 of 22/12/2000 (2000)

[2] EU. Directive 2007/60/EC of the European Parliament and of the Council of 23 October 2007 on the assessment and management of flood risks, European Parliament, Council, 2007

[3] Barbour M, Gerritsen J, Snyder BD and Stribling JB 1999 Rapid Bioassessment Protocols for use in streams and wadeable rivers: periphyton benthic macroinvertebrates, and fish (Washington DC: US Environmental Protection Agency – Office of Water) p 337

[4] Rinaldi M., Surian N., Comiti F., Bussettini M. (2016). IDRAIM – Sistema di valutazione idromorfologica, analisi e monitoraggio dei corsi d'acqua – Versione aggiornata 2016 – ISPRA – Manuali e Linee Guida 131/2016. Roma, gennaio 2016.
Barbour M, Gerritsen J, Snyder BD and Stribling JB 1999 Rapid Bioassessment Protocols for use in streams and wadeable rivers: periphyton benthic macroinvertebrates, and fish (Washington DC: US Environmental Protection Agency – Office of Water) p 337

[5] Buffagni A, Demartini D and Terranova L 2013 Manuale di applicazione del metodo CARAVAGGIO - Guida al rilevamento e alla descrizione degli habitat fluviali (Roma: Istituto di Ricerca Sulle Acque) p 293

[6] Kleynhans CJ, Louw MD, Thirion C, Rossouw NJ and Rowntree K 2005 RiverEcoClassification: manual for EcoStatus determination (Pretoria: Joint Water Research Commission and Department of Water Affairs and Forestry) p 210

[7] Raven PJ, Holmes NTH, Dawson FH, Fox PJA, Everard M, Fozzard IR and Rouen KJ 1998 River Habitat Quality: the physical character of rivers and streams in the UK and Isle of Man (Bristol :Environmental Agency) p 86

[8] Siligardi M, Avolio F, Baldaccini G, Benabei S, Bucci MS, Cappelletti C, Chierici E, Ciutti F, Floris B, Franceschini A, Mancini L, Minciardi MR, Monauni C, Negri P, Pineschi G, Pozzi S, Rossi GL, Sansoni G, Spaggiari R, Tamburro C, Zanetti M 2007 IFF 2007 – Indice di Funzionalità Fluviale (Trento: Agenzia Nazionale per la Protezione dell'Ambiente e per i Servizi Tecnici) p 336

[9] Ristorto, G., D'Incalci, P., Gallo, R., Mazzetto, F., & Guglieri, G. 2017 Mission Planning for the Estimation of the Field Coverage of Unmanned Aerial Systems in Monitoring Mission in Precision Farming. Chemical Engineering Transactions, 58, 649-654.

[10] ENAC, 2018. "Remotely Piloted Aerial Vehicles" Issue 2, Revision 4. 21st May 2018.
Yellowscan, 2018. https://www.yellowscan-lidar

IOP Conf. Series: Earth and Environmental Science **275** (2019) 012006 doi:10.1088/1755-1315/275/1/012006

Measurement and estimation of leaf area index (LAI) using commercial instruments and smartphone-based systems

Raffaele Casa[a], Deepak Upreti[b], Federico Pelosi[c]

Department of Agriculture and Forestry Sciences, University of Tuscia, Viterbo, Italy.

[a]rcasa@unitus.it, [b]dupreti@unitus.it, [c]federico.pelosi.84@gmail.com

Abstract. The experimentation carried out within this project aimed at comparing various methods used to estimate leaf area index (LAI), which is a key variable to describe the vigour and productive potential of a crop. The direct or "destructive" method requires the removal of the biomass of the sample and is a very laborious and burdensome technique. The indirect estimates consist in the acquisition of a series of readings taken directly in the field with specific instruments, mainly based on the measurement of the quantity of light intercepted by the foliage and on the use of complex mathematical models. Indirect methods allow for a quick estimation of the LAI and there is a strong interest in developing economic and user-friendly tools. The tools compared in the present study included two state-of-the-art, rather expensive, commercial instruments, i.e. the Sunscan ceptometer (Delta-T, UK) and the LAI-2000 (Li-Cor, USA). These were compared to inexpensive alternatives offered by hemispheric photography from a smartphone equipped with a fish-eye lens and from a smartphone app called "Pocket LAI". All the methods tested showed the ability to provide an effective estimate of the LAI even if sometimes with an underestimation of the value measured by the direct method.

1. Introduction

Leaf area index (LAI) is a key biophysical variable for both crop monitoring and modelling applications, defined as the total one sided leaf area in relation to the ground [1]. It has been widely used in different agricultural applications [2] and remote sensing studies [3–5] including precision agriculture [6]. It is used also as key state variable in many crop growth models [7].

There are two approaches for measuring LAI in the field : direct and indirect methods [8, 9]. The direct or destructive method involves the removal of the biomass of the sample that is intended to be measured and is a very laborious and costly method. The indirect estimation consists in the acquisition of a series of readings carried out directly in the field with specific instruments, mainly based on the measure of the quantity of light intercepted by the foliage and the use of complex mathematical models. These commercial instruments estimate LAI from the gap fractions [10] measured from different angles, which is the fraction of sky seen from below the canopy (upward photography) or the fraction of soil seen from above the canopy (downward photography) [11].

There are different commercial instruments available for the indirect estimation of the LAI, and have good potential to estimate LAI as compared to direct destructive methods in many experimental conditions [12]. Sensors like LAI-2000 Plant Canopy Analyzers (LI-COR, Inc., Nebraska, USA) and DHP are some of the most widely used classical optical instruments for indirect LAI estimation [10, 13–15].The downside of these instruments is that they are quite expensive, both in the terms of purchase and maintenance. An alternative to these instruments is the use of digital hemispherical photographs acquired with a camera by attaching a fish eye lens to it. This method requires generally professsional (and expensive) cameras and lenses and large processing time for image classification after acquiring images to obtain a good LAI of the representative sample. This limits the use of these instruments in real time [11].

As an alternative, exploiting the features available in the smartphones, a mobile application called PocketLAI was developed [16] to estimate LAI. This app was tested on paddy rice against commercial instruments LAI-2000 and Decagon Accupar ceptometer. LAI estimation is based on the segmentation of images acquired at 57°, below the canopy to estimate gap fractions [13]. It proved its reliability in terms of both trueness and precision.

Since all the methods for indirect estimation of LAI depend on the models for light transmittance into the canopy and these models are strongly affected by canopy architectures [9, 17], testing these methods on canopies with different structures is mandatory. The aim of this study was to compare LAI estimation from smartphone app, PocketLAI, DHP. LAI-2000 and Sunscan Ceptometer. The reasons behind this study were the encouraging results obtained from [16] and the need to test them on other crops as well.

2. Materials and Methods

The measurements with the different methods was carried out in 2014 and 2015 on a total of 39 samples. The measures took place at the Maccarese SpA farm (17 samples) in the province of Rome (41.83° lat. N, 12.22° long. E, alt. 8 m a.s.l.). The surveys carried out for the other 22 samples took place at the experimental farm of the University of Tuscia, located in Viterbo (42.72° lat. N, 12.12° long. E, altitude 310 m). The measures were carried out on: spelt (*Triticum dicoccum* L.), durum wheat (*Triticum durum* Desf.), maize (Zea mays L.), alfalfa (*Medicago sativa* L.) and broad beans (*Vicia faba* ssp. *minor* L.).

2.1. Instruments Used

In this study direct, destructive LAI estimates have been considered as a reference LAI estimate and other indirect methods of LAI estimation are compared with this reference. Indirect methods of LAI estimation are – Sunscan Ceptometer (Delta-T, UK), LAI-2000 Plant Canopy Analyzer (LI-COR, USA), Hemispherical Photography from smartphone camera (equipped with fish eye lens), and smartphone app (PocketLAI).

2.2. Sampling Protocol

Many factors can influence the measurement technique used for a given coverage, such as height, size, sky conditions and sensor field of view [18]. For this reason, the measures were carried out with each instrument following a precise sampling protocol. The data of the same intruments were acquired in crop areas within frames placed on the ground of 0.9 by 1 m size.

The measurements were carried out in different plots of cultivated fields, choosing portions of the field in which the crop was homogeneously distributed and had an average height of about 30 cm. The frames were placed on the ground, thus separating the sample from the remaining crop. It was taken care in the choice of the exposure in the light of the selected surface unit, to ensure that the light conditions were such as to minimize the percentage of error in the readings.

2.2.1. Direct Methods

The LI-3100C (LI-COR, Lincoln, USA), area meter was used for the direct (or destructive) estimation of the LAI. The device is designed to be able to carry out a precise measurement of the foliar area, for leaves of any size. The leaves are collected from the total sample and placed one by one on a transparent conveyor belt that allows them to pass in front of a linear scanner.

The leaves travel under a light source so that the projected image is subsequently reflected by a three-mirror system on a scanning camera. The measurement of foliar area that derives from it is shown on the display, allowing reading and is expressed in cm^2. The roller, of which the instrument is fitted, is used to crush the leaves on the conveyor belt to avoid an underestimate of the real value, due to the foliar flap crumple.

With regards to the protocol used for the direct method of the LAI estimation, it was necessary, first, to remove the sample biomass, inside the quadrat, cutting the plants at the base of the collar, taking care not to harvest plant litter that could cause LAI overestimates. The collected biomass was weighted to extract sub-samples equal to about one third of the sample weight. The sub-sample, built from whole plants, was subdivided dividing the leaves from the branches and weighted them separately. Then both the leaves and stems were passed through the meter area. The value estimate by the meter area is considered as the one closest to the real value and therefor plays a fundamental role in comparing different methods.

2.2.2. Indirect Methods

2.2.2.1. LAI-2000

The data acquired with the LAI-2000 instrument were collected by taking a reading above the canopy and six readings below the canopy for each quadrat. In order to carry out the six readings under the vegetation, inside the quadrat, the sampling scheme shown in Fig.1a was adopted. The spatial layout of the readings was conceived in order to carry out an estimate of the LAI as much as possible corresponding to the sample of the plant cover contained within the quadrat, subsequently collected for the destructive determination by the direct method. A limited field of view was adopted up to the fourth ring (for zenith angles 0° to 58°), and a cover to limit the circular acquisition sector to 90°. The area measured by the instrument, for each reading, was calculated using:

$$r = \tan(58) * H \quad (1)$$

$$A = \frac{\pi r^2}{4} \quad (2)$$

where H is the height of the vegetation. In order to correct for the error introduced because the measurements were not acquired under diffuse light conditions, a calibration of the instrument was carried out according the LAI-2200 manual (LI-COR, Lincoln, USA). A series of readings were acquired with the use of a diffuser cap: 1) with diffuser resting on the sensor in the shadow, 2) with diffuser resting on the sensor exposed to direct light, 3) with the sensor without the diffuser, in the shadow. Also, maximum attention was paid to ensure that the sensor was shaded, in order to avoid exposure of the sensor to direct sunlight.

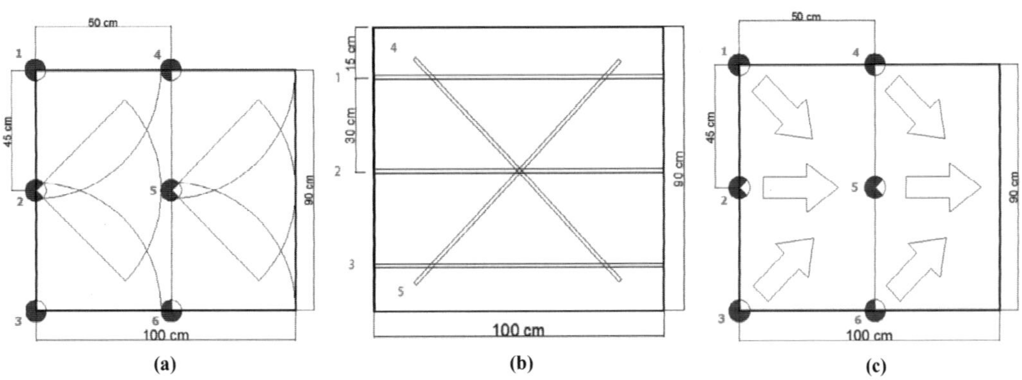

Fig1. Sampling Plan - a) LAI-2000 b) Sunscan Ceptometer c) PocketLAI

2.2.2.2. SunScan Ceptometer

The sampling protocol used for Sunscan ceptometer consisted of five readings carried out under the canopy cover and one above it *[19].* The measurements were obtained, for the first three readings, by placing the rod on three points, equidistant from them, parallel to the greater side of the quadrat (Fig. 1b). The fourth and fifth readings were made by positioning the rod at 45°, with respect to the vertices of the quadrat, and the last reading was done above the crop to measure the incident radiation. Finally, an additional reading was carried out to measure the radiation reflected from the soil on a adjacent area where there was no vegetation.

2.2.2.3. Digital hemispherical photograpy

For the methodology based on hemispheric photography an IPhone-5 (Apple Inc, USA) equipped with a special cover was used. A fish eye lens suitable for the camera of the smartphone was employed. The readings consisted in the acquisition of a series of hemispherical images taken by placing the instruments on the ground with the lens of the camera oriented towards the sky. The camera has a field of view limited to 50° with respect to zenith and a cover was used that limited the acquisition to a circular sector of 90° to allow to mask the operator from the photo. After checking the contrast between the cover and the sky, the images were stored and subsequently processed by the CanEye software [19] up to the LAI estimation. The measurement protocol used for this method is the same that is used for LAI-2000 Fig1a.

2.2.2.4. PocketLAI – Smartphone App

A Samsung Galaxy Xcover GT-S7710 smartphone with an Android 4.1.2 operating system was used to test the PocketLAI app. Once, the app is started, the smartphone is in read mode and the operator has 5 seconds to position the device below the vegetation. Shortly, after positioning the smartphone, the application emits a first signal of vibration, the orientation of the screen is continuously recalculated using the accelerometer sensor. When the angle between the vertical and the normal reaches 57.5°, the image is captured and processed, using an algorithm described by [15]. Similarly, further readings are calculated, and a LAI value is estimated. The protocol used with PocketLAI application is shown in Fig1c.

3. Results and Discussions

Table 1 shows the comparison between the values of LAI obtained through the destructive sampling and those acquired by indirect methods. Comparisons include all data collected during the period from 10 March 2014 to 28 January 2015. The range of LAI values explored in this study is reasonably wide to the values reported in [19–21]. Fig. 2 presents scatterplots showing R^2 and RMSE values.

Digital hemispherical photographs captured by Iphone with fish eye lens attached provides the best estimates of LAI for alfalfa (RMSE = 0.33 m^2/m^2) followed by PocketLAI (RMSE = 0.40 m^2/m^2) (Table 1).

Table1. RMSE values of LAI estimates for different crops and instruments

	RMSE (m^2/m^2)			
	Sunscan Ceptometer	**LAI-2000**	**DHP Processed**	**PocketLAI**
Alfalfa	0.88	0.60	0.33	0.40
Broad bean	0.30	0.48	0.30	0.65
Durum wheat	1.17	0.83	0.87	1.05
Emmer	1.54	1.16	0.31	1.77
Maize	0.87	0.89	0.58	0.77
Wheat	1.36	1.05	0.68	1.46

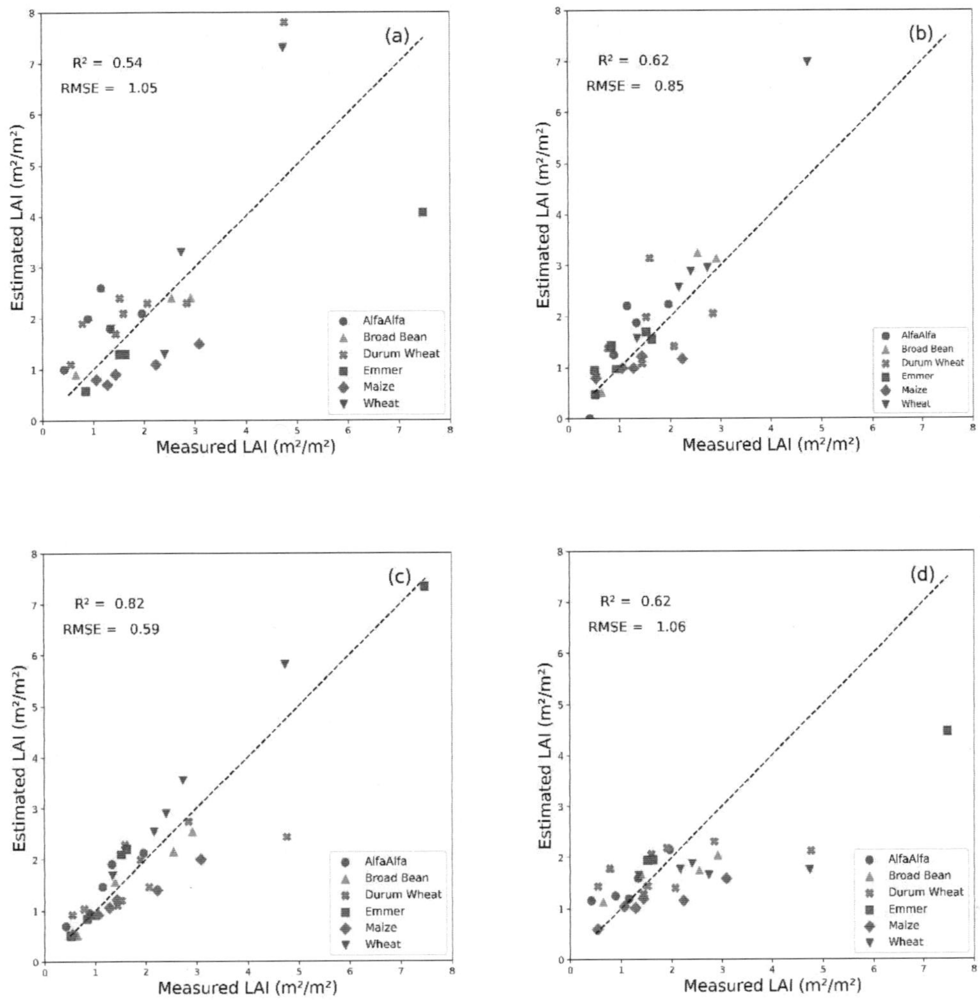

Fig 2. R^2 and RMSE for a) Sunscan Ceptometer, b) LAI-2000, c) DHP d) PocketLAI Application

For broad bean, LAI lowest error was estimated by Sunscan Ceptometer (RMSE = 0.30 m^2/m^2) and DHP (RMSE = 0.30 m^2/m^2). In case of durum wheat LAI values estimated were best achieved using LAI-2000 (RMSE = 0.83 m^2/m^2) and DHP (RMSE = 0.87 m^2/m^2), higher values of LAI showed some over estimation for wheat with SunScan Ceptometer. DHP images resulted in the lowest error in emmer (RMSE = 0.31 m^2/m^2), maize (RMSE = 0.58 m^2/m^2) and wheat (RMSE = 0.68 m^2/m^2). PocketLAI mobile application although have satisfactory performance, it showed some under estimation and saturation effect for LAI values higher than 2.

In this study, DHP processed with the CanEye image processing software obtained the best overall results in terms of RMSE. The downside of this method is the time taken for processing the images after acquisition. It is impractical for real time deployment. Commercially available instruments LAI-2000 and Sunscan Ceptometer are although widely used for the indirect estimation of LAI and accepted globally as a reliable source to estimate LAI, these instruments are expensive in terms of cost and maintenance. PocketLAI application for LAI estimation was found satisfactory in terms of goodness of fit and error estimates, it is an economic and feasible alternative for large and frequent field campaigns. As, the PocketLAI application used in this work calculates gap fraction in only one incident direction (~ 57.5^0), some modifications of the calculation of gap fraction, by for example considering gap fractions from multiple directions and taking an average of them, may lead to better and improved estimation of LAI.

4. Conclusions

All the methods showed the ability to provide an estimate of the LAI that is closer to the estimation of the measurement with the area meter, that is the direct reference measure, the more the canopy meets the assumptions based on a random distribution of the leaves. In general, with regard to the accuracy of the instrument, the performance of the app was very similar to that of LAI-2000. A better performance of the LAI-2000 was found as compared to Ceptometer. As has already been observed by many authors, all methods adopted for the estimation of LAI generally lead to an underestimation of the measured value to the direct method [7, 8]. In this study some over estimation was observed with digital hemispherical photographs, LAI-2000 and Sunscan Ceptometer (Fig. 2). The main problem encountered with LAI-2000 was an underestimation due to a saturation phenomenon, which was also found in Sunscan Ceptometer, which occurs under very high crop density conditions. Removing the external ring redings slightly more reliable. Regarding the PocketLAI app, a underestimation for LAI values greater than 2 was found. However, this method has undisputed advantages over all other instruments, especially as regards the greater practicality of use and the economic aspect: a few euros in case the users are already in possession of a smartphone, compared to approximately 10 thousand euros planned for the purchase of tools like LAI-2000 or Sunscan. An additional advantage of the app is the elimination of the costs of repair and maintenance, thus excluding the risk of a possible interruption of the measurement campaign in case of damage of the device. However, the problems shown by the app cannot be attributed exclusively to the calculation processes it works to estimate the value of the foliar area index because they derive it mostly from the field vision (58°) of the camera used for the measurement. The performance of the hemispherical photography-based method was the best in terms of accuracy among the methods used. Also, in this case it has been shown that it is possible to use low cost equipment, consisting of a smartphone and a fish eye lens purchased online for a few euros. However, this method requires a later phase of processing the images on computers, using specific software such as CanEye, not allowing an instantaneous calculation of the LAI and requiring a specialist competence in the data processing phase.

References

[1] Chen JM, Black T. Defining leaf area index for non-flat leaves. *Plant Cell Environ* 1992; 15: 421–429.

[2] Jay S, Maupas F, Bendoula R, et al. Retrieving LAI, chlorophyll and nitrogen contents in sugar beet crops from multi-angular optical remote sensing: Comparison of vegetation indices and PROSAIL inversion for field phenotyping. *Field Crops Res* 2017; 210: 33–46.

[3] Richter K, Hank TB, Vuolo F, et al. Optimal exploitation of the Sentinel-2 spectral capabilities for crop leaf area index mapping. *Remote Sens* 2012; 4: 561–582.

[4] Vuolo F, Neugebauer N, Bolognesi SF, et al. Estimation of leaf area index using DEIMOS-1 data: Application and transferability of a semi-empirical relationship between two agricultural areas. *Remote Sens* 2013; 5: 1274–1291.

[5] Upreti D, Huang W, Kong W, et al. A Comparison of Hybrid Machine Learning Algorithms for the Retrieval of Wheat Biophysical Variables from Sentinel-2. *Remote Sens* 2019; 11: 481.

[6] Haboudane D, Miller JR, Pattey E, et al. Hyperspectral vegetation indices and novel algorithms for predicting green LAI of crop canopies: Modeling and validation in the context of precision agriculture. *Remote Sens Environ* 2004; 90: 337–352.

[7] Jin X, Kumar L, Li Z, et al. A review of data assimilation of remote sensing and crop models. *Eur J Agron* 2018; 92: 141–152.

[8] Bréda NJ. Ground-based measurements of leaf area index: a review of methods, instruments and current controversies. *J Exp Bot* 2003; 54: 2403–2417.

[9] Weiss M, Baret F, Smith G, et al. Review of methods for in situ leaf area index (LAI) determination: Part II. Estimation of LAI, errors and sampling. *Agric For Meteorol* 2004; 121: 37–53.

[10] Ross J. The radiation regime and the architecture of plant stands, The Hague: Junk W.

[11] Campos-Taberner M, García-Haro FJ, Confalonieri R, et al. Multitemporal monitoring of plant area index in the Valencia rice district with PocketLAI. *Remote Sens* 2016; 8: 202.

[12] Verger A, Martínez B, Camacho-de Coca F, et al. Accuracy assessment of fraction of vegetation cover and leaf area index estimates from pragmatic methods in a cropland area. *Int J Remote Sens* 2009; 30: 2685–2704.

[13] Wilson JW. Estimation of foliage denseness and foliage angle by inclined point quadrats. *Aust J Bot* 1963; 11: 95–105.

[14] Jonckheere I, Fleck S, Nackaerts K, et al. Review of methods for in situ leaf area index determination: Part I. Theories, sensors and hemispherical photography. *Agric For Meteorol* 2004; 121: 19–35.

[15] Garrigues S, Shabanov N, Swanson K, et al. Intercomparison and sensitivity analysis of Leaf Area Index retrievals from LAI-2000, AccuPAR, and digital hemispherical photography over croplands. *Agric For Meteorol* 2008; 148: 1193–1209.

[16] Confalonieri R, Foi M, Casa R, et al. Development of an app for estimating leaf area index using a smartphone. Trueness and precision determination and comparison with other indirect methods. *Comput Electron Agric* 2013; 96: 67–74.

[17] Stroppiana D, Boschetti M, Confalonieri R, et al. Evaluation of LAI-2000 for leaf area index monitoring in paddy rice. *Field Crops Res* 2006; 99: 167–170.

[18] Biosciences L. Lai 2000 Plant Canopy Analyzer. Operating Manual. *Linc USA*.

[19] Facchi A, Baroni G, Boschetti M, et al. Comparing optical and direct methods for leafarea index determination in a maize crop. *J Agric Eng* 2010; 41: 33–40.

[20] Darvishzadeh R, Skidmore A, Schlerf M, et al. Inversion of a radiative transfer model for estimating vegetation LAI and chlorophyll in a heterogeneous grassland. *Remote Sens Environ* 2008; 112: 2592–2604.

IOP Conf. Series: Earth and Environmental Science **275** (2019) 012006 doi:10.1088/1755-1315/275/1/012006

[21] Ceotto E, Di Candilo M, Castelli F, et al. Comparing solar radiation interception and use efficiency for the energy crops giant reed (Arundo donax L.) and sweet sorghum (Sorghum bicolor L. Moench). *Field Crops Res* 2013; 149: 159–166.

A full integrated system for agroclimatic and pest monitoring at farm and landscape scales in Campania Region

Raffaele Martino[2], Massimo Nicolazzo[2] and Giuliano Langella[1]

[1] Department of Agriculture,
University of Naples Federico II, Via Università 100, 80055 Portici, NA, Italy
[2] Department of Electrical Engineering and Information Technology,
University of Naples Federico II, Via Claudio 21, 80125 Naples, NA, Italy

E-mail: {raffaele.martino2,nicolazzo,glangella}@unina.it

Abstract. The current agrometereological monitoring of the Campania Region is both inadequate for its users and insufficient to support the simulation of pest risk models. This work describes a cyber-physical system which has been designed to overcome these limitations, with the objective of (i) automate all the tasks involved in climatic data management, including also the intervention of the human expert when appropriate, (ii) develop dependable pest risk models and provide the input data they need, and (iii) provide real-time agrometereological data presentation. The system is built around an automatic climatic data management engine called WeatherProg, which manages data collection, quality control, data reconstruction, and digital maps production. A proper climatic database has been developed to support the operations of WeatherProg, as well as a Web application for real-time publication of data handled by the software. Pest risk models, fed by point measurements and the digital maps produced by WeatherProg, are going to be prototyped. Furthermore, a new small measurement station has been prototyped, with the aim of being low-cost and easy to relocate, in order to support the characterisation and models parameterisation for different areas. This paper illustrates the current status of the system and discusses future directions.

1. Introduction

The agrometereological monitoring system of the Campania Region currently comprises about 35 agrometerolgical stations covering the whole territory. The technical staff in charge of the network manages by hand collected data, and publishes daily aggregations on weekly basis, although the measurements are performed at a 10-minutes time scale.

This solution is clearly unsuitable for the needs of the operators in the Campania Region, which range from agronomy professionals who support the farmers, to phytopatological experts in charge of producing pest risk bulletins for the agricultural farms in the Region. The limited number of the available stations cannot properly capture and monitor the complexity of the regional environment. Moreover, the manual management prevents the possibility of real-time delivery of agrometereological data.

To face all these problems, a collaboration was established including an agrometeorological and pest monitoring work package within the URCoFi Executive Plan (Unit of coordination and strengthening of the activities of surveillance, research, experimentation, monitoring and education in the phytosanitary field), which involves the Department of Agriculture of the University of Naples Federico II and the Phytosanitary Central Office of the Campania Region, amongst other partners. This work package includes also the development and execution of pest risk models for the different agroecosystems.

This work presents a cyber-physical infrastructure which has been developed to accomplish all these tasks. The core of the whole system is the WeatherProg software [1, 2], a computer program capable of performing all steps of data management, from their collection to the production of digital maps of the climatic variables of interest, which in turn are used to feed the pest risk models. Data is currently collected from the existing institutional network, but in order to monitor and characterize the microclimatic environments of the fields in which pests will be studied and monitored, a new small, low-cost, and low power consumption measurement station has been prototyped. This station has been designed to be easy to relocate, in order to enable the characterisation of pests and risk models on different areas, where different crops are produced. Finally, the data processed by the system are made available to the public by means of a web application, specifically designed to allow users to easily retrieve data of their interest.

It is worth mentioning that the human expertise is not cut off by this cyber-phisical infrastructure. In fact, WeatherProg is intended to be a semi-automatic software: for instance, the ultimate decision about anomalous data is left to the intervention of the regional officers. In order to allow them to easily interact with the backend software, and to tune its parameters, a web dashboard has been designed and is going to be implemented.

2. Materials and Methods

2.1. The study area

Located in southern Italy between 13°45′E, 15°49′E, 39°59′N and 41°31′N, Campania is the third most populated region of the country, but due to its extension of about 13 600 m^2, it is the most densely populated region of Italy. Its inland is occupied by the Apenine Mountains, oriented roughly NW-SE, whereas the Sele and Campana plains border the coast. The Sele plain is named after the river which traverse it, while the Campana plan is traversed by the Volturno river. Nevertheless, the majority of the regional territory is hilly, for the 50% of the territory, against the 35% mountainous and the 15% flat. Consequently, the elevation ranges from 0 m to 1904 m above mean sea level.

The agrometereological monitoring is in charge of the Agrometereological Regional Centre (CAR) under the Department for Agriculture of the Campania Region. The station network run by the CAR includes, at the time of this writing, 34 stations spanning an elevation range from 11 m to 794 m a.m.s.l.

Not only the number of available stations is too limited to properly capture the aforementioned complexity of the regional environment, but also the equipment of these stations is highly variable. All the stations measure air temperature, rainfall and air humidity, but for 10 stations these are the only climatic parameters measured. Air pressure is measured by only 9 stations. All the 11 fundamental climatic variables are measured by at least one station in the network, but no one station is equipped to measure all the climatic variables.

2.2. The cyber-physical infrastructure

The architecture of the proposed cyber-physical infrastructure is illustrated in figure 1. The sensor networks on the territory provide input data to the system. Raw data are collected by the CAR, which sends them to the WeatherProg engine on hourly basis with a lag of one hour from the real time measurement. However, the format of the received data is not suitable for subsequent climatic operations, hence the first task of the WeatherProg software is to decode the received data, and store it in a WMO-compliant climatic data base, developed using the PostgreSQL relational DBMS [3].

IOP Conf. Series: Earth and Environmental Science **275** (2019) 012007 doi:10.1088/1755-1315/275/1/012007

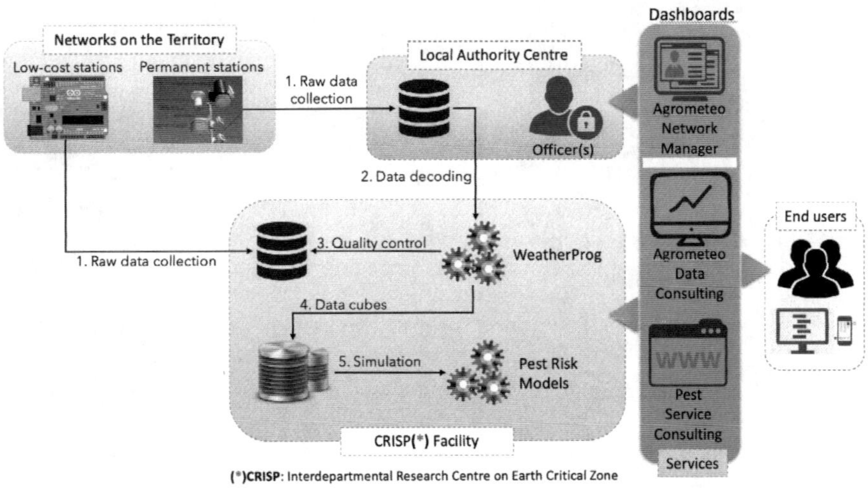

Figure 1. High level architecture of the proposed cyber-physical system

WeatherProg then performs a range of quality controls to spot and flag anomalous measurements, and reconstructs both wrong and missing data. This step purposedly involves some manual intervention from an human expert, in order to finally decide from the wrongness of anomalous data. In fact, apart from clear-cut correct and clear-cut wrong masurements, there is a range of suspicious cases which may or may not be correct. In such cases, rather than trying to make a guess, the measurement is flagged as suspicious and left to the human expertise for the ultimate decision. For the case study described in this paper, the human experts are the regional officers of the CAR. A Web-based dashboard allows the expert to review the output of the quality control and make the ultimate decisions. Moreover, the dashboard allows also to configure some of the parameters of the models embedded in WeatherProg.

The following step of the climate data management chain performed by WeatherProg is the production of digital climatic maps of relevant variables. According to the needs of the pest risk models which will consume them, digital maps of both fundamental variables like temperature or rainfall, and derived variables like evapotranspiration, can be produced by WeatherProg.

Digital maps are stored in a raster database called *rasdaman* (*Raster Data Manager*) [4], which makes available the datacube representation. A *datacube* is a representation of a digital map with more than two dimensions, where the additional dimensions can encode additional information. The simplest case is a three-dimensional map where the x and y dimensions represent easting and northing as usual, while the z dimension represents the time of the digital map. This datacube can actually be regarded as an "array of digital maps", where each element of the array is a digital map at a specific point in time. Example of datacubes are shown in figure 2a.

Once datacubes have been created by WeatherProg and inserted into rasdaman, pest risk models can be run, representing the last link of the chain. Thanks to the datacubes, a mechanistic pest model can be run on whatever point within the study area. Moreover, a pest risk map may also be produced by running the model over an area of interest. Figure 2b graphically explains how operations on datacubes work.

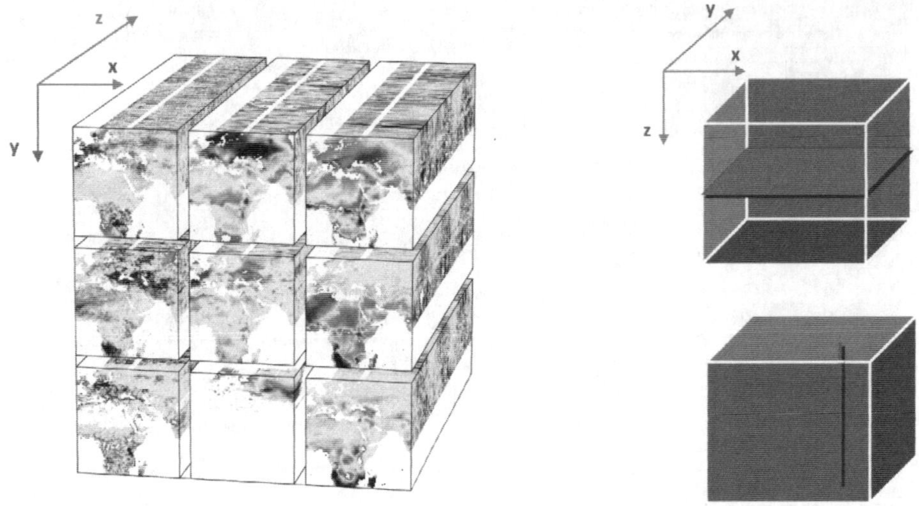

(a) Digital maps varying in time, represented as datacubes

(b) Operations on a datacube

Figure 2. Datacube examples. Note that y and z axes are swapped between (a) and (b)

3. Results

3.1. Main Engine

The WeatherProg software was originally developed within the SOILCONS-WEB project (LIFE08 ENV/IT/000408). It requires some modifications in order to be inserted within the cyber-physical infrastructure proposed in this work, and to become capable of working with data at the time scale of 10 minutes.

The decoding module of WeatherProg has been completed. This module interacts with the CAR data servers on one hand, which send the data of the regional network, and the climatic data base on the other hand. Since the format of incoming data is subject to change, and has actually changed during the period of this study, the interface between WeatherProg and the incoming CAR data has been made flexible, so so as to be capable of handling future format modifications effortlessly.

The same flexibility has been implemented for the interface between WeatherProg and the climatic data base. To this end, instead of directly making SQL queries wherever needed by the various WeatherProg modules, the software makes appropriate function calls to a dedicated module, which in turn takes in charge to appropriatly querying the database.

3.2. Low-cost Stations

The design of the low-cost measurement station has been driven by a twofold objective. Firstly, the coverage of the various regional environments by the CAR network (only 34 stations) is numerically insufficient and geospatially too coarse in order to build adequate digital maps that can support the prediction of accurate agrometeorological time series at ungauged geospatial locations. Lastly, these stations must be cheaply maintained and easily relocable over the years, in order to monitor different areas and/or agroecosystems. It is crucial also considering the possibility by a farmer to invest few moneys to get the standard equipment (hardware and software directly connected to our servers) we developed. This way the farmer can run pest risk models using in-situ measurements instead of those coming from predicted climatic cubes.

Pest risk models, in fact, requires a very precise characterisation of the environments at the farm scale in order to tune model parameters; but it is infeasible to have the required number

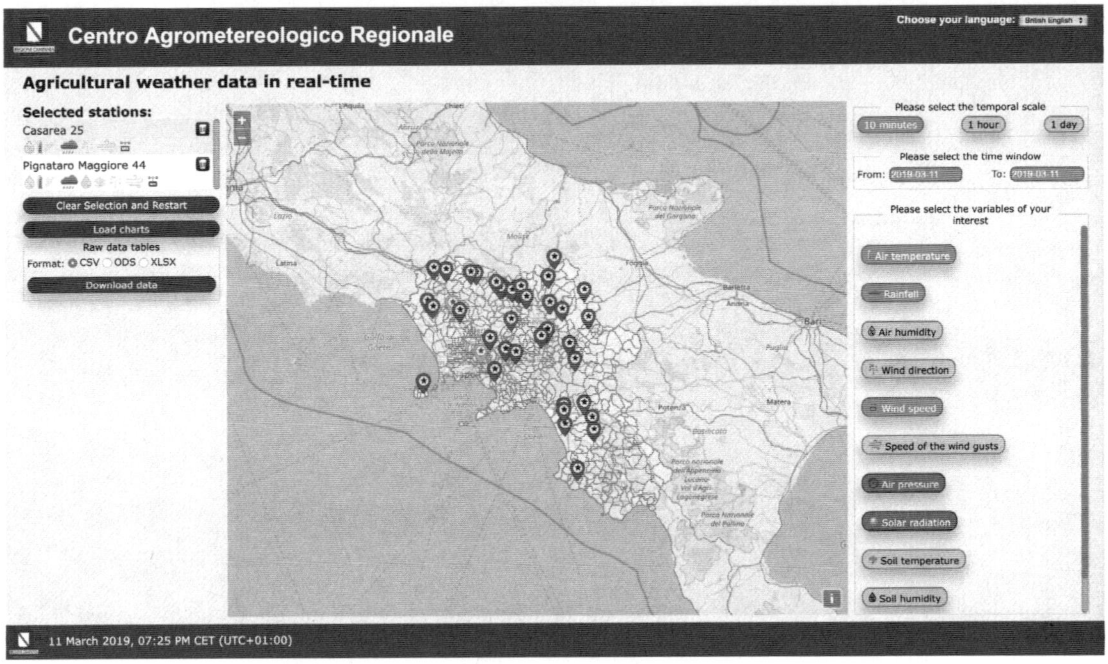

Figure 3. Web application for data presentation

of stations all across the regional territory. Alternatively, a reduced number of stations can be deployed in a limited area, sufficiently for a proper characterisation of the pest risk models, and being moved after the model parameters have been tuned.

The measurement station prototype is based on an Arduino-like platform, embedding an Espressif microcontroller [5] and a Sigfox transmission chip [6]. It has been developed to be energy-independent, in order to work with little or no manutention effort. Moreover, it has a good computational performance, and is capable of retrieving, hold when needed, and, transmit information from a wide range of sensors.

Each component of the measurement device has been chosen according to its cheapness, but also its dependability, durability and, for the sensors, accuracy. This choice has led to the best trade-off between affordability and quality for the device as a whole.

The fact that the resulting equipment is very low cost makes this solution appealing for a multi-point farm monitoring. The device is currently equipped and programmed to work with sensors measuring air temperature, precipitation, leaf wetness, relative humidity and wind direction, speed and gusts, but other sensors can be easily added and monitored so as to expand the set of measured climatic parameters.

3.3. Web Applications

An easy-to-use, GIS-like web application has been developed to allow end users to retrieve data of their interest. It is shown in figure 3.

The interface guides the user towards the selection of the available data. Upon loading, the user is asked for the selection of one or more stations of the network. The selection of the climatic variable is enabled only when at least one among the selected stations is capable of measuring that climatic variable. Moreover, the appropriate aggregation for each climatic variable is displayed when a time scale higher than 10 minutes is selected.

The Web application then produces a chart for each requested climatic variable, showing

IOP Conf. Series: Earth and Environmental Science **275** (2019) 012007 doi:10.1088/1755-1315/275/1/012007

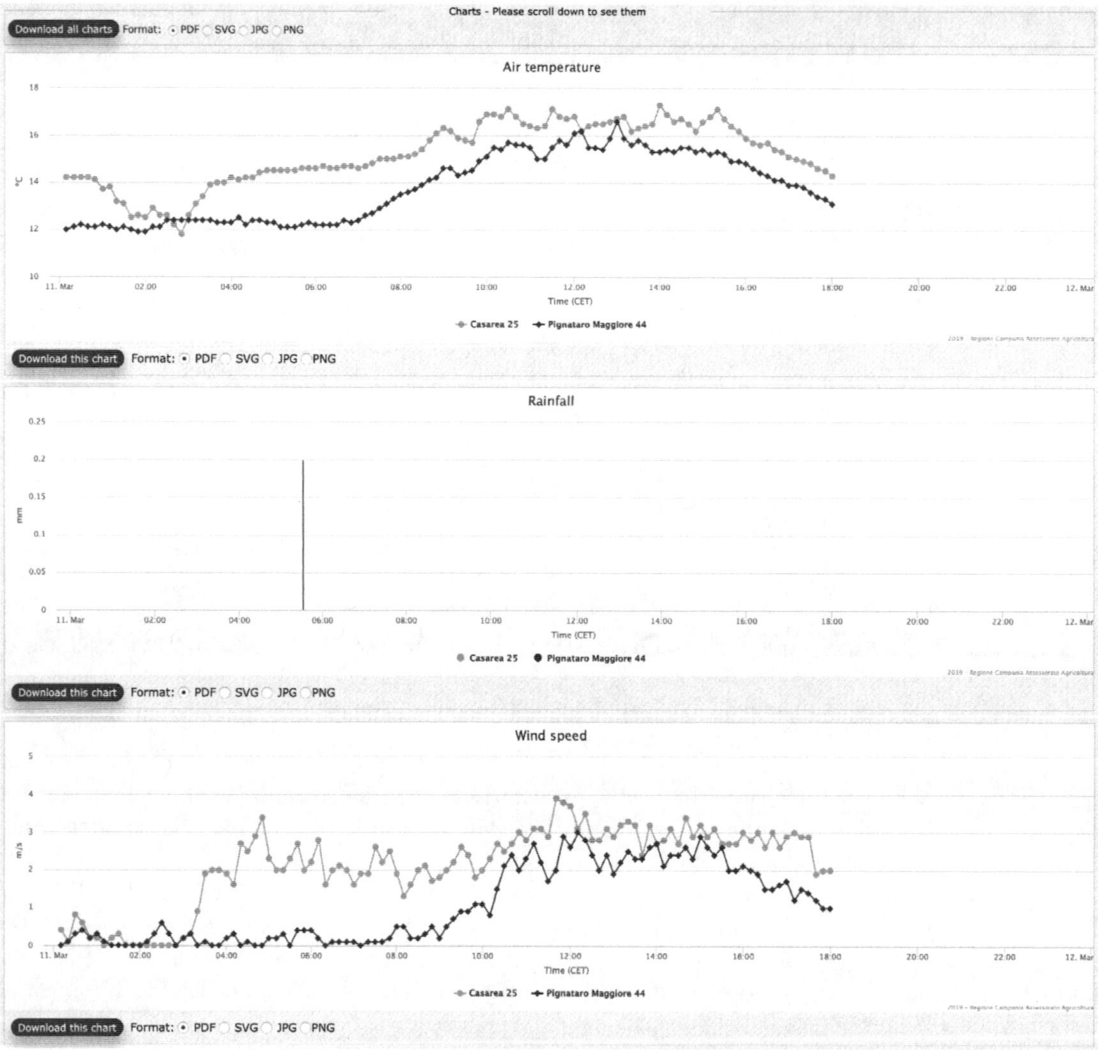

Figure 4. Example of a search result

the evolution of the climatic parameter during the time interval of interest. An example of the produced charts is shown in figure 4. Both charts and data in a spreadsheet can be downloaded, in a variety of formats including also open formats for spreadsheets.

4. Discussion

The work discussed hitherto proves the feasibility and the advantages of a semi-automatic management and real-time publication of the climatic data collected by the institutional agrometereological network of the Campania Region. However, this is only a starting point. The work outlined here has, in fact, laid the foundations for the complete end-to-end cyber-physical infrastructure.

4.1. Main Engine

The quality control logic has already been updated to work at the 10 minutes time scale. However, while some checks can be performed on the data point in isolation, like the logical checks, there are also more complex checks like the spatial and climatological ones, which require a quality-controlled time series to be used as reference. Therefore, the quality control will be enabled by firstly perform the simpler checks on all the available data, and then the more complex modules will be enabled.

4.2. Low-Cost Stations

The development of the low-cost station requires the evaluation of the accuracy of the measurements. This analysis will be performed shortly after the first deployment of the devices. The evaluation will take into account the presence of the quality control performed by WeatherProg, in order to assess the dependability if the whole cyber-physical system.

Moreover, the low-cost stations will be relocated on a yearly basis, in order to study and optimize the parameterization of the models for different combinations of crops and pests.

4.3. Web Application

The Web application will be expanded to present also data from the low-cost devices. The user will be able to choose whether to visualize the data of the institutional network, the low-cost station network or both.

Data presentation will be further expanded with OGC-compliant data delivery mechanisms, especially for raster data, which will be added when available from the rest of the chain.

5. Conclusions

The proposed integrated approach has proven capable to overcome the limitations of the current agrometereological monitoring system, allowing for real-time quality control and data presentation which would not be possible with the current manual-only data management.

However, the importance of the human expertise is recognised, as the system is not built as a closed, unmodifiable package which produces ultimate decisions. Instead, WeatherProg is intended to be a semi-automatic system, aiming to ease the work of the human experts by focusing their attention only on the critical pieces of data which require to be reviewed.

Apart from already providing value for the end-users and the regional officers, the results obtained by the implemented infrastructure strongly encourage the development of other functionalities of the system, which will result in a full integrated system for agroclimatic and pest monitoring.

Acknowledgments

This work has been funded by the Department of Agriculture of the Campania Region, within the URCoFi Executive Plan 2018-2019.

References

[1] Langella G 2014 Weatherprog: a computer program for the automatic handling of climatic data *Role of agrometeorology in the new politics in agriculture* p 102

[2] Langella G 2016 Semi-automatic handling of meteorological ground measurements using weatherprog: prospects and practical implications *European Geosciences Union (EGU) General Assembly, 17/22 April 2016, Vienna, Austria*

[3] The PostgreSQL Global Development Group 2019 Postgresql: The world's most advanced open source relational database URL `https://www.postgresql.org`

[4] Baumann P 2018 rasdaman - raster data manager URL `www.rasdaman.org`

[5] Sparkfun ESP32 Thing URL `https://www.sparkfun.com/products/13907`

[6] Sigfox Breakout board BRKWS01
URL `https://partners.sigfox.com/products/sigfox-breakout-board-brkws01`

IOP Conf. Series: Earth and Environmental Science **275** (2019) 012008 doi:10.1088/1755-1315/275/1/012008

Proposal of an ontological approach to design and analyse farm information systems to support Precision Agriculture techniques

F Mazzetto[1,2], R Gallo[1], M Riedl[2], P Sacco[2]

[1] Faculty of Science and Technology, Free Univ. of Bozen/Bolzano (FUB), P.zza Universitá 5, 39100 Bolzano (IT)
[2] Fraunhofer Italia, Via Volta 13A, 39100 Bozen/Bolzano (IT)

E-mail: *fabrizio.mazzetto@unibz.it*

Abstract. After a comparative evaluation around the concepts of Industry 4.0, Precision Agriculture and Smart Farming, the paper discusses the necessity of identifying solutions able to apply the methods of the so-called Knowledge Management 4.0 in the agri-environmental enterprises. To this aim, an ontology based on a conceptual map derived from the data-to-information transformation is here proposed to support the design of a new generation of Farm Information Systems (FIS) able to fit the production needs of agri-environmental enterprises. The data-to-information cycle is split into four phases: A) Data collection, B) Data processing, C) Data analysis and evaluation, and D) Use of information. Phases A and D comprise tools of "light digitization" typically referring to OLTP components (On Line Transactional Processes). On their turn, phases B and C are typically formed by "heavy digitization" components (On Line Analytical Processes), generally more complicated to be managed by the farmers directly. The conceptual map defined by the ontology is firstly useful to plan the composition of the FIS according to a modular approach, possibly following a strategy that starts to consider the introduction of OLTP components, for then evolving gradually towards more articulated solutions that include also OLAP component and related functions. In this way, decision makers can manage practical planning instrument able to show clearly the level of complexity a FIS can reach, thus evaluating its sustainability with respect to the financial, cultural and professional resources available at the farm. Finally, after having underlined the opportunities offered by Cloud Computing and widespread connectivity, that provide an easier adoption of OLAP tools, some application cases are presented and discussed.

1. General background

1.1. Industry vs. Agriculture

The recent proposals of the so-called *Industry 4.0* (**Ind-4.0**) have introduced new paradigms in the management of production processes through ICT [1, 7]. Many solutions can also be applied in the agricultural sector, within which the application of *Precision Agriculture* (PA) techniques has been considered for over twenty years [2, 3, 4].
An exhaustive definition provided by the US NRC (1997, [8]) states that PA is a "management strategy that uses information & communication technologies (ICT) to collect data from multiple sources in view of their later use in decisions concerning production activities". Originally, this definition was firstly intended to refer to field processes, being mainly focused on the highly automated site-specific

IOP Conf. Series: Earth and Environmental Science **275** (2019) 012008 doi:10.1088/1755-1315/275/1/012008

approaches, aiming at overcome the management limits imposed by the relevant spatial variability in field properties (from which also the names of "*prescription-* " or "*target-farming*"). Later, it was extended to many other types of farming systems, such as livestock, viticulture and orchards. Despite the relevance of the original definition, for a long time the PA has been mainly intended as a means to allow the transfer of advanced automated applications into the agricultural sector. This created great confusion on the market among potential users, often disappointed by completely unsatisfied expectations. Only recently also the name "*Smart Agriculture*" has started to be used by many producers in the sector, probably more to reflect a fashion rather than to affirm a new technological principle.

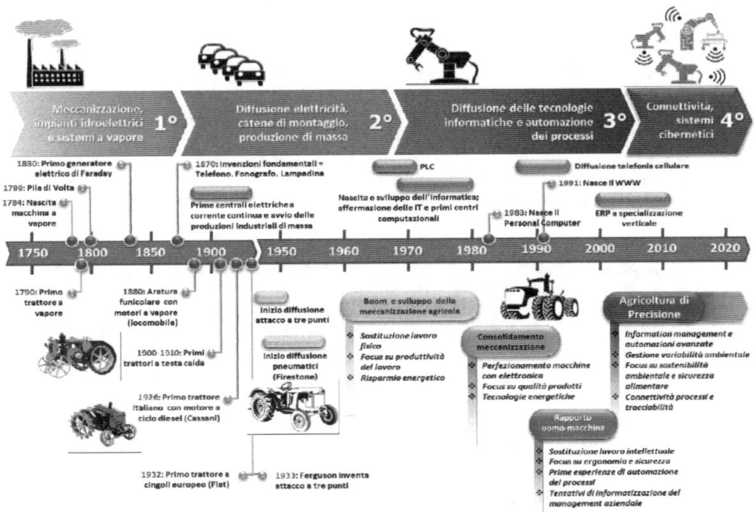

Figure 1. Comparative scheme of the evolutionary stages of industry and agriculture

Surely this term has its usefulness because it tends more to shift the attention on the need to realize *forms of quality management* rather than pursuing mere purposes of advanced automation. In relation to this, the quality of management can be firstly intended as the ability to make decisions based on targeted information, previously collected through a global monitoring of production processes, thus reconnecting to the original NRC´s definition. Sometimes even the name "Agriculture 4.0" is provided, just to underline an assonance of concepts with Ind.4.0. However, as shown in Fig.1, this is rather improper because the 4th industrial revolution implicitly mentioned in the concept of Ind.4.0 follows the 3rd industrial revolution approximately occurred at the end of last century (1970-2000). Basically, this is the period of the industrial digitization with the introduction of **ERPs** (*Enterprise Resource Planning*) and related information systems. This has not occurred in agriculture and the technological gap to be tackled must also take into account the shortcomings in the use of ICT. Here, therefore, every innovation must consider different approaches, starting from the proposal of new farm information systems (FIS) able to manage the problems of agro-environmental enterprises whose production means are delocalized in the territory and subject to specific needs as indicated in Fig.2 (moving processes in uncontrolled open spaces, climatic aleatory, lower professional skills, multitasking workers, scarce financial resources) [5, 6].

In summary, with regard to the five different aspects considered in Fig.2, we note that: 1) industry and agriculture have diametrically opposed tendencies; all aspects that are generally favourable to the industry represent serious weaknesses for agriculture; 2) there are actually different types of agriculture; the most negative effects relate to arable farms and crop cultivations, while animal husbandry shows less suffering due to production methods typical of industrial processes (obviously except for forage production processes); this also explains why the first experiences of digitization of agriculture were related to the management of milking processes; 3) the sector of farm contractors shows further strengths, both for the high specialization in technologies, often equipped with advanced automated solutions, and for a better skill of employees, basically more suited to pay more attention to the quality of management (although also here, obviously, the weak point of the management of the processes on large portions of the territory remains still relevant).

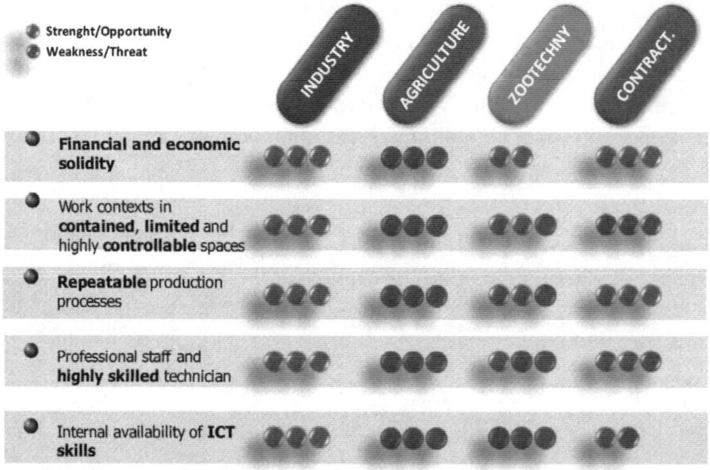

Figure 2. Aspects influencing ICT use in different production contexts

1.2. Industry 4.0 & Precision Agriculture

Ind-4.0 and PA share many objectives, such as: 1) process digitization, necessary for confirming and reinforcing the use of *Information System* for managing production processes, planning resources and supporting advanced strategic decisions at the enterprise (ERP evolution); 2) searching for *robotics* and *advance automated processes* (for both monitoring the production lines and improving process productivity); 3) *hyper-connectivity* between production agents (both human and machines), searching for advanced cybernetic approaches (for the moment, with a large focus on industry, only); 4) Treatment of huge amount of data (*Big Data*), and 5) *Machine Learning*, with quick data interpretation to make information easier to use in decision-making processes.

Internet of Things (**IoT**) and *Internet of Services* (**IoS**) are tools and methods that can largely contribute in supporting good results for the above objectives. As mentioned, these objectives are much more feasible for the industry than for agriculture, since the latter has yet to recover a significant technological gap on the ICT front. In fact, here the digitalisation of the sector has been lacking, as basically has been limited to the diffusion of electronic technologies (basic automation) on board the machines (especially tractors).

Nowadays, however, even the introduction of automated solutions always requires the use of digital applications designed to become increasingly complex. In this regard, we can speak of "*light digitization*", when ICT requires simple firmware embedded in the application devices, designed to convert in real time the signals measured by sensors or identification systems into elementary actions through appropriate actuators. Instead, we can speak of "*heavy digitization*" when the measured data are recorded on special supports to be used in later times, after appropriate processing, transformation and evaluation. From this point of view, we can state that while in the Ind.3.0 phase the industrial sector (in parallel with the commercial sector) has contributed to consolidate technological solutions, with related professional profiles, suitable for managing forms of heavy digitization, in the agricultural sector even today many users interested to PA tend to focus more on solutions of light digitization, showing considerable inertia in moving towards mixed forms.

1.3. Knowledge Management 4.0 (KM4.0)

Smart Agriculture, similar to Ind.4.0, has the objective of implementing the quality production systems by applying forms of information management in which *light + heavy digitization components are integrated*. This is possible through the introduction of Information Systems (IS) that provide tools to support the transformation of raw data into information. The latter in turn are then transformed into decisions through the so-called Knowledge Management (**KM**) phases described in Fig.3. Raw data represent a message of the real world that is worth registering when a priori is expected to be of some importance in the decisions to be taken with respect to the farm domain of interest. Data becomes information when actually enter the decision-making process. The decision maker then manipulates the information transforming them through his cognitive processes first into "Knowledge" (apply context) and then into "Wisdom" (apply insight), to finally arrive to the actual decision (apply purpose).

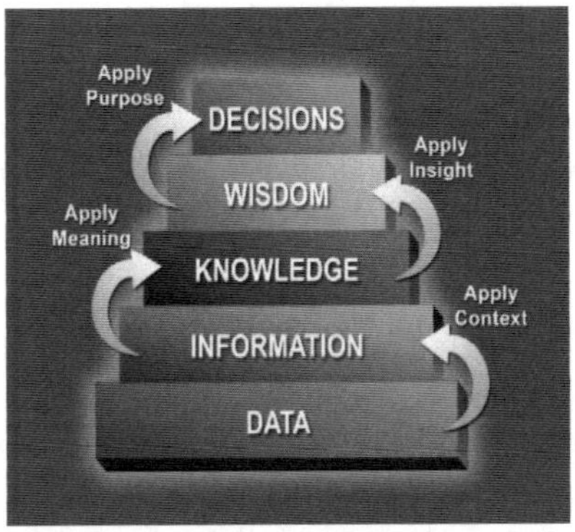

Figure 3. Transformation of raw data into information and then into decisions according to the view of Knowledge Management

Therefore: 1) the IS, with all its components, supports the transformation *"raw data into information"*; 2) the Decision-maker then personally performs the transformation *"information into decision"* throughout his own mental processes.
In short, ISs enable decision makers to:
a) Collect data related to processes, materials and products;
b) Process and archive the collected data, through models and databases coherent with the future decisions to be taken;
c) Analyse and use processed data, being supported in the transformation of information into knowledge and wisdom, according to their decision-making processes;
d) Access information at any time they need.

Thus, the IS becomes the main KM instrument, since it also promotes the sharing of knowledge among all the members of the productive system, each with his/her own *hierarchical role* (decision-making level). From this standpoint, KM is a practice independent of current technological levels as it tends to pursue a sort of "philosophy" of collaboration and sharing in the workplace.
Thus, the concept of knowledge based mainly on the skills of individuals is overcome (while is still strongly common in traditional agricultural contexts), to affirm instead the need to produce new knowledge only through the sharing and processing of information. Obviously, the ICTs have largely contributed to affirm this conceptual model, so much so that the vision of Ind.4.0 parallels the concept of **KM4.0**. The essential aspects of the latter include:

- Ensuring an approach highly oriented to **IoT** knowledge processes;
- Managing and treating Big Data acquired directly from **things** (= *elements of processes and products*) and **customers** (*people acting in the system*);
- **Sharing information** between people or things without any limitation;
- Storing all data and information directly in **clouds (IoS);**
- Ensuring that all contents are always available **online**, also to implement any **real-time** automation process;
- Providing information sharing (***C2C, C2M***) via wireless solutions (**Hyper-connectivity**);
- Fostering of **predictive analysis** in the main maintenance and control tasks.

Even Smart Agriculture must be able to guarantee a gradual process of methodological and technological evolution along the direction of these approaches.

2. Scope

An ontology is here proposed to design Farm Information Systems (**FIS**) specifically conceived for agro-environmental enterprises. The latter, as we have seen, present profound differences with respect to industrial companies, above all in terms of size, profitability, staff and work organization, need to manage mobile processes distributed throughout the territory with all the related implications (necessity of positioning, adaptation to weather conditions, site-specific variability of lands, need to process and monitor biological and environmental elements).

In industrial companies, classic ERP systems are generally divided into 5 vertical sectors (1. human resources, 2. warehouse, 3. production, 4. logistics, 5. accountability), which are integrated within an IS of varying complexity. Regardless of the type of enterprise, all sectors are usually of equal importance in the IS structure, and the system tends to maintain a certain degree of management autonomy in each sector. Things change significantly in the Ind.4.0 modalities, with an increase in the relative importance of the production and logistics sectors when compared to the other three.

In the case of KM4.0 applied to agro-environmental enterprises, the FIS can generally be limited to only three vertical sectors, namely:

1. **Warehouse**, to keep track of all the materials in input and output for the company, including energy items, thereby also responding to traceability tasks;
2. **Processes**, with the functions of: a) monitoring, which here assumes - also for environmental constraints - fundamental and often cogent importance, b) synthesis and analysis, with the support of a series of packages for specific decision support to be evaluated from case to case; c) control, with functions of direct intervention on the individual processes through sensors or actuators (automation);
3. **Administration**, including staff, customers, suppliers and accounting.

The need of an ontology for FIS designing is here felt, as we have to deal with: i) an application domain that cannot be clearly split into independent sectors, ii) a core "process domain" that must share many common resources among many production goals, iii) the need of limiting to internal re-organization of the farm as a consequence of the introduction of new managing tools, both for cultural and financial constraints.

The ontology aims to provide a conceptual map capable of describing components and functionalities of the FISes themselves, defining either the strategies for their modular development, or the level of management and structural complexity to be achieved. The "map" starts from the conceptual difference existing between "data" and "information" and takes into account the whole pathway with which data are transformed into information.

3. Approach and methodology

3.1. OLTP and OLAP components

The life cycle data-to-information is described in Fig.4.

The cycle includes 4 phases: **A.** *Data collection*, **B.** *Processing*, **C.** *Analysis and evaluation*, **D.** *Use*. In each phase the data take different states: raw data (A), data processed (B), data evaluated or formatted (C), information (D).

The scheme also distinguishes between 2 FIS areas featured by different levels of digitalization:

- The **operational areas**, referring to specific elements of production processes, dealing with individual events with great details and usually following highly standardized procedures; they include components that have in charge single, specific transaction, and are even named **OLTP** components (*On Line Transactional Processes*); they include the phases A and D of the data-to-information cycle;
- The **informational areas**, with opposite characteristics, as include components dealing with aggregated facts through overall views, analyses and global controls typically carried out through flexible procedures in which the experience and the subjective standpoint of the various decision-makers often weigh heavily; here the key-point is a global analysis approach and they are even named **OLAP** components (*On Line Analytical Processes*); they include the phases B and C of the data-to-information cycle.

IOP Conf. Series: Earth and Environmental Science **275** (2019) 012008 doi:10.1088/1755-1315/275/1/012008

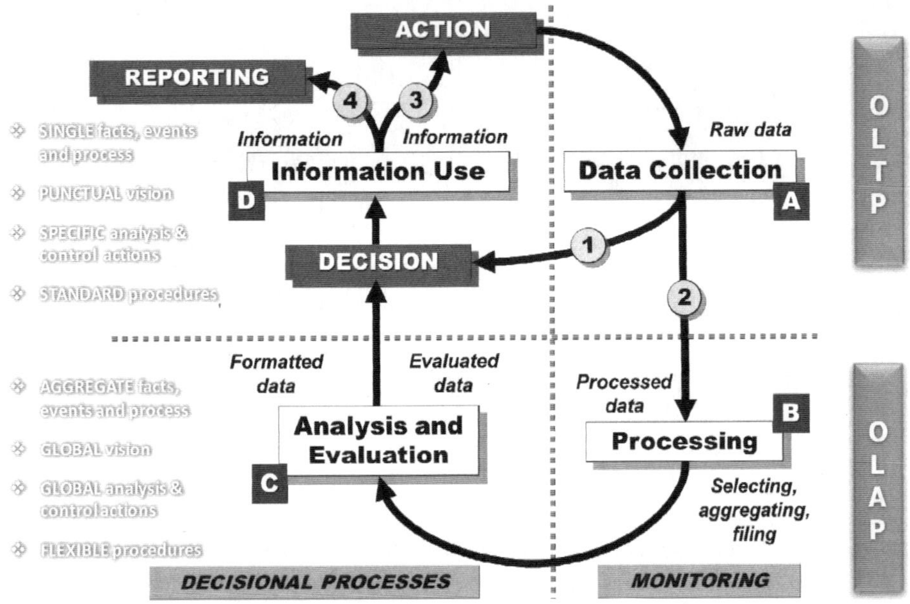

Figure 4. Schematic representation of the data-information life cycle. The scheme highlights the information flows and the conceptual methods of integration between operational components (**OLTP**, *On Line Transactional Processes*) and informational components (**OLAP**, *On Line Analytical Processes*)

The above features make OLAP components significantly more complex to use than OLTP components. *That's way OLTP and OLAP components can be assimilated to the afore mentioned concepts of light and heavy digitization solutions, respectively.*

3.2. FIS phases

Phase A consists in the acquisition of data of potential interest to the decision maker. The acquired data can then undertake two alternative routes:

1. be *used immediately*, even through real-time processing, to make instant decisions (automated processes);
2. provide for *deferred use over time*, thus passing to **phase B**, with the archiving and possible integration/synthesis with other data from different periods; it is the path of common management applications, with the typical directional control functions.

The cycle closes with the **phase C** related to analysis and **phase D** related to the use of the information. The decisions to be taken concern problems of:

3. *control of on-farm processes*, operational or managerial; the decision is followed by an action to intervene on the processes themselves;
4. *certification towards third parties* (verify of production protocols or product quality standards); the decision then becomes a documentation activity (traceability).

In each phase, the FIS must provide a variety of hardware and software components, to support (or completely replace) the work of decision-makers, within the phase. To this aim we can distinguish between aiding tools and automating devices. For this reason, the scheme also becomes useful for defining a "*conceptual map*" with which to identify the different components of the FIS to: i) facilitate the design of the FIS; ii) provide adequate forms of integration between components; iii) define its degree of complexity to evaluate a priori the difficulties of its transfer in the production context.

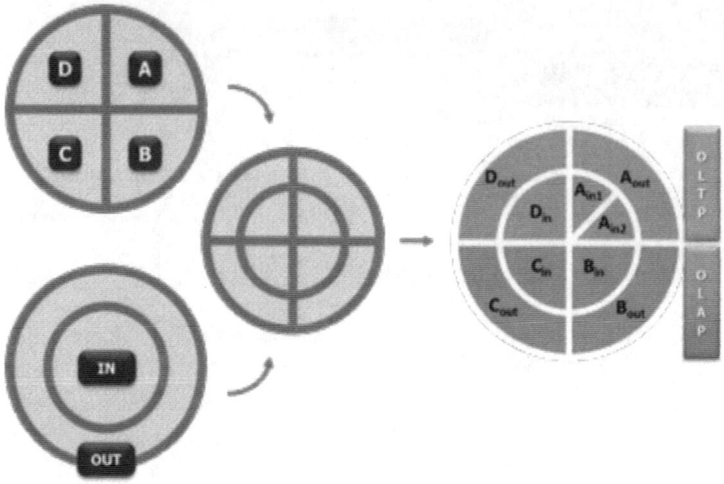

Figure 5. Approach derived from previous Fig.4 to define a conceptual map with which to classify the components of an information system. For internal data collection, two situations are taken into account to distinguish between crop monitoring (**A_{IN1}**) and operational monitoring (**A_{IN2}**)

Such a map can be derived from Fig.4 according to the approaches presented in Fig.5, where we observe:

- four circular sectors corresponding to the four afore mentioned functional FIS phases;
- two concentric sectors **(IN, OUT)** related to the degree of integration of potential FIS components with the enterprise's core business, as well as with the FIS data warehouse.

By overlaying circular and concentric sectors, we obtain the final conceptual map of the FIS, in which each *X-th* phase is divided into two distinct X-IN and X-OUT fields. "IN" sectors indicate aspects intimately connected to the execution and management of farm internal processes, with the implication that here every FIS component must be necessarily "customized" on the characteristics of the farm structures and behaviour, usually with direct connections to the central FIS DBMS. "OUT" sectors, on their turn, indicate components linked to situations that are more conditioned by the different external environment types with which the farm interacts (meteorological, physical, commercial, regulatory, etc.). Their components are limited to forms of conceptual integration, with limited or completely absent connections to the central DBMS.

3.3. FIS sectors

The FIS sectors that can be thus identified on the conceptual map are defined as follow:

- **A_{IN1}: Production monitoring**, including any component aiming to achieve data on conditions and *properties of elements that are the objects of the production process*; in case of cultivations, it is referred as "*crop monitoring*", and for example includes all sensors used to detect NDVI or yield mapping data;
- **A_{IN2}: Operational monitoring,** including any component aiming to achieve data on the *ways by which a process is carried out*; frequently, it requires the application of identification system for the automatic detections of all the agents that take part to a process (power units, implements, workplaces etc.);
- **A_{OUT}: Environmental monitoring,** it usually aims at the observation of the physic-chemical variables that characterize the *environment in which the productive activity takes place*; in case of crop cultivation, it frequently concerns the use of sensors for the measurement of pedological and meteorological parameters;
- **B_{IN}**: **Basic processing**, it regards the use of essential ICT components for the *basic digitalisation* of management processes; therefore, in addition to the fundamental hardware (PCs, PDAs, any servers, network connections, etc.) and the operating system, it firstly includes the components

of the farm data warehouse, together with all the tools for modifying and querying the DBMS; secondly, there are the classic software packages for office automation (word processing, spreadsheets, graphics programs, e-mail, etc.);

- **B_{OUT}: Advanced processing**, it provides the use of software packages of medium and high complexity, normally dedicated to the *management of digital maps* interfaced with DBMS (GIS), possibly also supported by advanced management procedures of vector graphics (CAD), geo-referencing, image analysis and statistical processing and / or geostatistics. Their use is typically intended for expert users (knowledge workers), who in large enterprise should have to permanently assist the farm direction. Since this is usually not possible, most of the features of this phase are often outsourced;
- **C_{IN}: Customized assessments,** they concern the use of *Decision or Management Support Systems* (DSS and MSS) directly integrated in the FIS structure for an efficient and direct data exchange with the farm DBMS; these are often highly complex applications based on stochastic simulation methods that require knowledge of historical and climate local data. As such, they operate on mainly structured or semi-structured information. Their use is occasional or relatively frequent and require the assistance of highly qualified personnel (knowledge workers or external technical assistance); for example, they could concern programs for: the definition of fertilization plans, the creation of prescription maps in chemical spreading operations, the simulation of the productive behavior of alternative cultivation plans; the creation of diagnostic maps etc .;
- **C_{OUT}: General utility assessments**, they concern *applications for general analyses*, often limited to a "conceptual" integration in the FIS, rarely used and mainly intended to operate on unstructured data, totally independent from the FIS DBMS, often with the need for access to external databases (open access applications available in the net); they usually relate to models based on deterministic calculations and optimization methods, such as packages for: exercise cost estimation, investment analysis, energy balances, LCA analysis (Life Cycle Assessment), multicriteria analysis etc.
- **D_{IN}: In-Farm use of information**: it usually concerns the implementation domain of production activities, and provides for the use of all those components of various kinds (frequently actuator devices, but also displays for improving the H2M communication) that allow the use of the information in farm processes featured by different levels of automated execution.
- **D_{OUT}: Out-Farm use of information**: it includes the set of tools able to make the farm suitable to produce the necessary documentation to perform, according to various purposes, *process and/or product certification tasks* (both on compulsory or voluntary basis); application examples can range from the simple printing of invoices or transport documents, till the formulation of complete EPD reports (Environmental Product Declaration); related components are usually represented by specific software packages suitable for this purpose; however, it still remains subordinated to the preventive implementation of both Monitoring (phase A) and Basic Processing (B_{IN}) phases.

3.4. Using the FIS conceptual map

The conceptual map of the FIS can firstly highlight the logic role of each hardware and/or software component within the FIS, according to the function it performs in the data-to-information life cycle. This also provides a prompt appraisal of the level of complexity at hand, according to the number and the distribution of the components: *higher the number of the components and the number of "activated areas", higher the FIS complexity*. As the application of OLAP components is generally more complex than OLTP ones, a step-by-step and modular introduction from OLTP to OLAP components should be always expected, in order to achieve a progressive perception of the usefulness from the FIS use, also becoming gradually more familiar in the conduction of information management finalized to a quality production system. As the application of OLAP components is generally more complex than OLTP ones, a step-by-step and modular introduction from OLTP to OLAP components should be always expected (Fig.6), in order to achieve a progressive perception of the usefulness from the FIS use, also becoming gradually more familiar in the conduction of information management finalized to a quality production system.

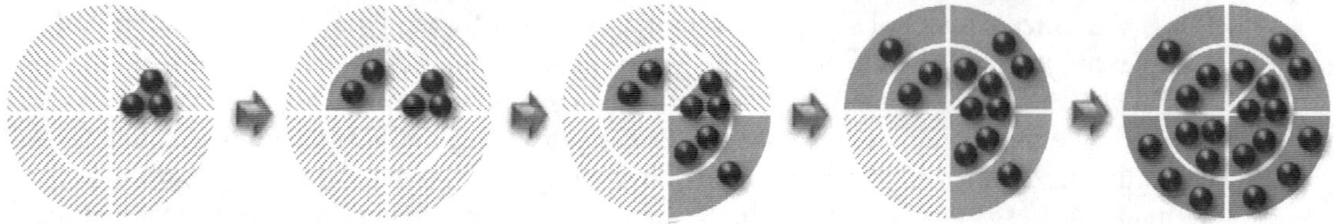

Figure 6. Qualitative example of a FIS evolution on a farm, starting from the introduction of OLTP and progressively moving towards OLAP solutions. Hardware (red) and Software components (blue) are here indicated.

The modularity should also ensure a strong control on investments, avoiding component redundancies or unnecessary repeated expenses. However, searching for modularity is still an hard task requiring the identification of standard approaches and methods that must pass necessarily through the agreement among the main player and producers of the sector.

4. Application examples

The proposed logical map offers a key to interpreting the levels of complexity and integration between FIS components in the design phase. It is not possible to define an "information system" as such if some OLAP components are not foreseen, starting from the presence of a DBMS, even in an essential format.

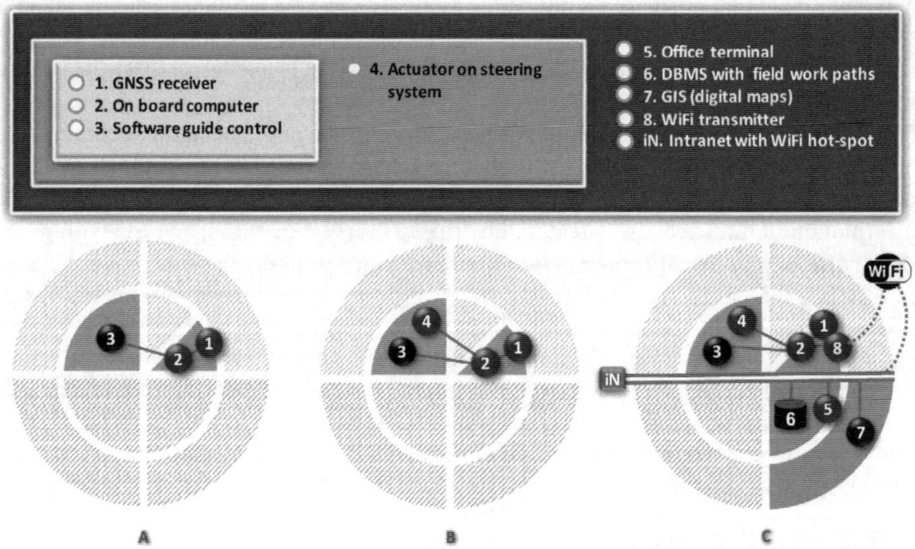

Figure 7. Example of classification of A FIS components according to the approach of Fig.6. A) Swath guidance (the driver is only assisted in the direction adjustments that must be manually provided); B) Automatic guidance (adjustments are automated by an actuator on the steer); C) As B, with in addition registration and consultation of the tracks.

A first example is given in Fig.7, where two solutions for assisted and automatic guidance are illustrated. In the first two cases (Figs.7A and 7B), we have stand-alone applications, without any connectivity with other users. However, the scheme highlights how even in this case it is advisable to proceed with possibly modular solutions: the automatic guidance is a sort of upgrade of the assisted one (even named

swath guidance), with the addition of an actuator device on the steering system, controlled directly by the same software already in use for the assisted driving.
Despite their simplicity, these applications have proved useful on several occasions as they have given rise to unprecedented interest in farmers in the use of IT, inducing in some of them the desire to also record the traces of the field operations for some later consultations. Although simple (but not trivial), this request introduces the use of OLAP systems (Fig.7C) because it requires the presence of a data storage system (B_{IN}) with related querying methods, possibly on GIS components (B_{OUT}). A further improvement implemented in Fig.7C is also linked to the introduction of a farm intranet line, to facilitate data transfer through a WiFi solution.
A further typical PA application example concerns the distribution of materials (pests or fertilizers) with automated dosages according to site-specific logics. Two approaches can be followed, depending on whether the dose control is performed by:

a) an *on-board sensor* (typically optical) that, based on predefined rules, allows to independently set the dosages in real time (**OLTP-oriented solution**, Fig.8);
b) a *pre-defined prescription map*, which combines the planned doses with the current position of the spreader (supplied by a GNSS device) to select the dose to be sent to a VRT system (**OLAP-oriented solution**, Fig.9A).

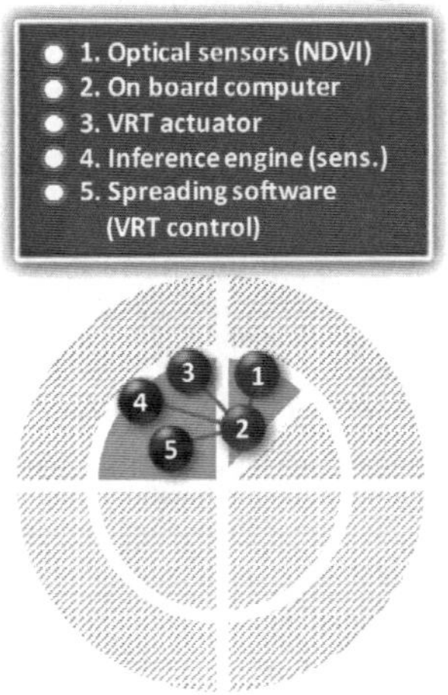

Figure 8. OLTP-oriented spreading solutions, driven in real time by tractor on board optical sensors.

Only this last solution needs to be supported by a FIS, as it requires the integration of synthetic analytical skills prior to the execution of the works. It could then evolve modularly towards even more integrated and complex solutions to include all the functional areas of the SIA, as indicated in Fig.9B. In particular, here we can observe: i) DSS for generating prescription maps; ii) environmental monitoring tools; iii) procedures enabling traceability reports. In addition, the increased complexity leads to the use of wireless data transmissions (GPRS) and the introduction of a Server with related Intranet.
In general, it is useful to keep in mind that:

- the higher the number of components in the OLAP phases, the more the management of the FIS will be complex, at least at an early stage;
- some OLTP components have no practical meaning unless properly integrated into specific OLAP components; for example, traceability functions require the presence of data storage systems that can be consulted on demand;
- the integration of systems within a FIS (i.e. solutions already composed of several tools) usually leads to positive synergistic effects in strengthening the analytical capabilities of decision makers. This is the case, for example, of the combination of automated electronic scouting solutions (crop monitoring) and operational monitoring in orchard farming systems or wineries: the integrated combination of knowledge of current activities and phytosanitary conditions significantly increases the information content finalized to diagnostic and management purposes.

5. Cloud computing and service centres

The applicative examples so far discussed, regardless of their level of complexity, are nevertheless confined within the framework of a "classical" digitization, not yet fully satisfying the requirements of the KM4.0 approaches. In fact, while ensuring a good degree of automation of spreading processes (Fig.9B), as well as automated monitoring activities supported by forms of wireless data transfer, hyper-connectivity levels are there still confined within the farm. In addition, the farm must be also provided with an internal server, with all the management complications of the case.

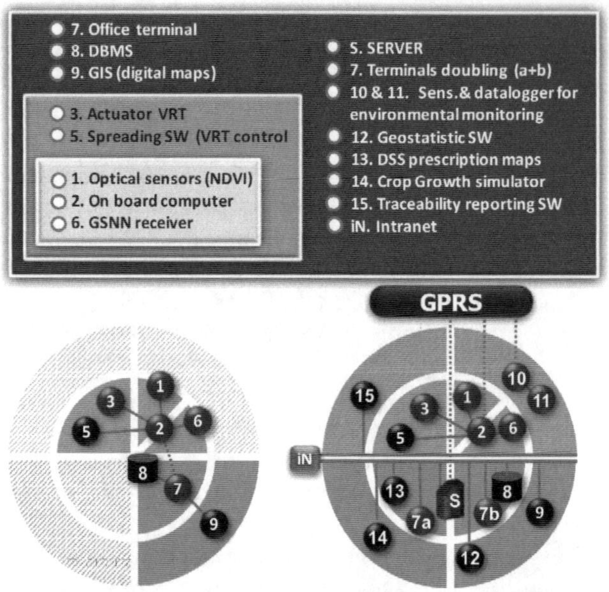

Figure 9. A) OLAP-oriented spreading solutions, driven in real time by a prescription map, manually predefined by the user (no MSS component in phase C); B) as A), but with extended functions related to data transmission (via GPRS and through an intranet line, with an internal server), environmental and operational monitoring, evaluations and traceability tools.

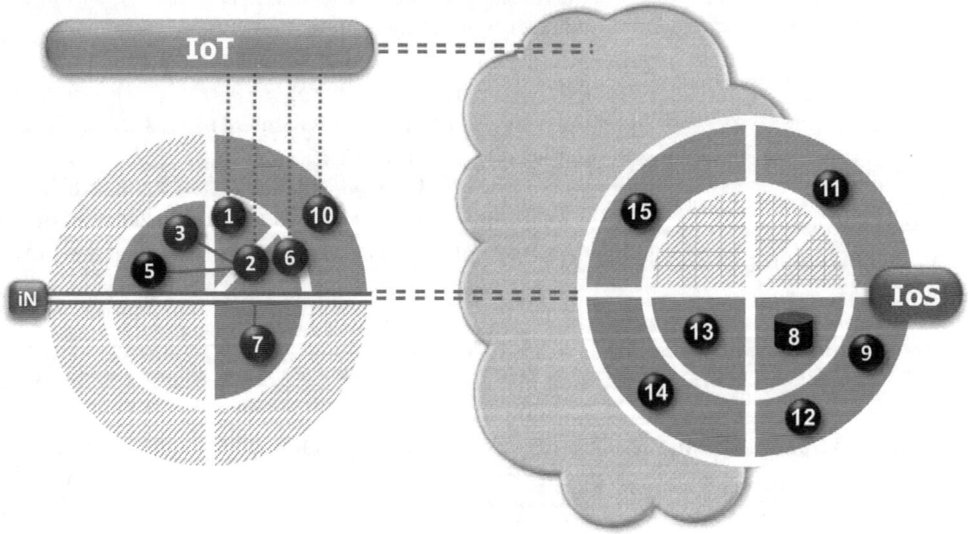

Figure 10. Evolution towards KM4.0 approaches of the previous solution discussed in Fig.9B. All the data and the main OLAP functions stay on the "Cloud" managed by a Service Centre.

A possible evolution towards a KM4.0 logic is described in Fig.10, in which we can note a sort of "FIS decoupling" where: a) a IoS platform enables to transfer all the OLAP complexities into the cloud (including the server maintenance); b) the most relevant OLTP components remain in charge to the farm, excepts the packages related to environmental monitoring and traceability; c) IoT solutions largely simplifies the hardware management of monitoring tasks; d) the main OLAP component that must remain in charge to the farm is a terminal for net connections with the IoS.

The new frontiers of Cloud Computing and widespread connectivity today open new possibilities for positive future developments of FIS KM4.0-based. Many commercial services are already available on the net, offering a large part of these FIS functions, making easier the availability of OLAP technological resources to remote users.

In general, there are three type of actors that have independent roles on the "cloud": 1) the *cloud provider*, which makes available under payment the required resources (virtual servers, archives, any possible applications); 2) the *administrator*, which configures the services offered by the cloud provider according to the final user's requests (he can also manage access rights and add any customized proprietary applications; 3) the *user* (final customer) who applies the services appropriately configured by the administrator.

It is desirable that as soon as possible new Service Centers specialized in Smart Agriculture (**SASC**) can be established and organized, in order to take in charge the roles of administrator, thus freeing the FIS responsible of single farms (users) - through totally online services - from the most burdensome responsibilities in the data management (physical maintenance of ITC equipment, with any intermediate data processing, including relative consistency checks) [4].

The support of the SASCs, moreover, could also favor the creation of diffused FIS, with a hierarchical over-company structure, in which single farms depend on a single main reference enterprise, enabled to perform control functions on single associate members. A typical example concerns the situation shown in Fig.11, in which N viticulture farms confer the grapes to a single winery. The latter has its own W-FIS which also performs monitoring tasks, aiming at controlling the quality of the supplied grapes through as much information as possible. These are obtainable from the F-FIS of the single conferring farms, available to share part of their DBMS with the winery. All roles are managed by the SASC, together with the data access policies compatible with the levels of confidentiality established between the parties. Here, too, all the main OLAP functions are left to the SASC, while all the peripheral farms remain focused on the management of the requested OLTP components.

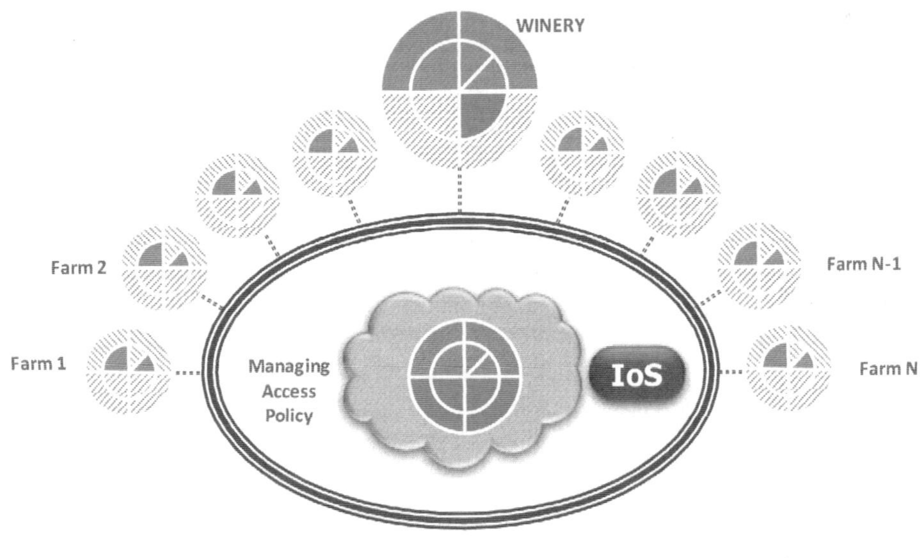

Figure 11. Conceptual example of a IoS platform providing FIS services to a winery together with all the farms conferring grapes to it (traceability of suppliers to improve grape selection for high quality wine production).

6. Final remarks

An ontology based on a conceptual map derived from the data-to-information transformation has been here proposed to support the design of a new generation of Farm Information Systems (FIS) able to fit the production needs of agri-environmental enterprises. Such a conceptual map is firstly useful to plan the composition of the FIS according to a modular approach, possibly following a strategy that starts to consider the introduction of OLTP components, for then evolving gradually towards more articulated solutions that include also OLAP component and functions. In this way, decision makers can manage practical planning instrument able to show clearly the level of complexity a FIS can reach, thus evaluating its sustainability with respect to the financial, cultural and professional resources available at the farm.

The FIS designing activity must be always organized following an infological approach (ie, oriented to the need of information), rather than a datalogical one (ie, oriented to the data that must be collected). Such a distinction is not trivial, and a mistake on this can occur very frequently (also in the better-trained industrial sector). Indeed, thinking of data is always simpler than thinking of problems to solve, on their turn linked to decisions that must be taken. The logical map provided by the ontology could be useful also to this aim.

Components selection, modular growths strategy and general FIS structure should be always decided trying to satisfy the requirements of a KM4.0 approach. To this aim the availability of a Service Center specialized in Smart Agriculture (SASC) is always a great help for the possibility of decoupling as much as possible OLTP and OLAP components at the farm, leaving the OLAP management directly to the SASC.

Moreover, these SASC could also provide on-site services, with supports for company auditing and training in the initial phase, as well as providing assistance in monitoring activities in the form of: i) rental of recording devices, in the case of operational monitoring (the ownership of the hardware components would thus remain in charge to the SASC, maintenance included); ii) execution of on-demand crop monitoring surveys with own means (with ground/remote sensing activities through terrestrial or aerial vectors, such as UAV); iii) processing and/or interpretation of user data.

The success of this approach will depend on: 1) SASC service reliability, with related completeness and timeliness; 2) SACS competence in identifying highly interdisciplinary forms of aggregation of their internal staff; 3) the ability to propose service tariffs compatible with the expected benefits in the use of FIS, not always easily to assess in purely monetary terms.

For sure, the needs of Smart Agriculture require to start a deep cultural revolution among its workers and responsible. While waiting for "digital natives", such a revolution must even start from revising the education programs in Agricultural Curricula to meet the digital perspective of future generations. In fact, while the management of the ITC devices to be used can stay in charge to engineer and computer experts, the management of knowledge and wisdom must remain in charge to agricultural experts, that however must be able to move with enough autonomy through the new technological frontiers of the sector.

References

[1] Aubert BA, Schroeder A, Grimaudo J 2012 IT as enabler of sustainable farming: An empirical analysis of farmers' adoption decision of precision agriculture technology. *Decision Support Systems* **54** pp 510-520

[2] Calcante A and Mazzetto F 2014 Design, development and evaluation of a wireless system for the automatic identification of implements. *Computers and Electronics in Agriculture* **101** pp 118-127

[3] Cox S 2002 Information technology: The global key to precision agriculture and sustainability. *Computers and Electronics in Agriculture* **36** pp 93-111

[4] Kaloxylos A, Groumas A, Sarris V, Katsikas L, Magdalinos P, Antoniou E, Politopoulou Z, and Maestre Terol C 2014 A cloud-based farm management system: Architecture and implementation. *Computers and Electronics in Agriculture* **100** pp 168-179

[5] Mazzetto F, Gallo R, Importuni P, Petrera S and Sacco P 2017 Automatic filling of field activities register, from challenge into reality. *Chemical Engineering Transactions* **58** pp 667-672

[6] Ristorto G, Mazzetto F, Guglieri G and Quagliotti F 2015 Monitoring performances and cost estimation of multirotor Unmanned Aerial Systems in precision farming. *2015 International Conference on Unmanned Aircraft Systems* ICUAS 2015 art. no. 7152329 pp 502-509

[7] Wan J, Cai H and Zhou K 2015 Industrie 4.0: Enabling technologies. *Proceedings of 2015 International Conference on Intelligent Computing and Internet of Things* ICIT 2015 art. no. 7111555 pp 135-140

[8] US NRC National Research Council 1997 Precision Agriculture in the 21st Century: geospatial and information technologies in crop management. *Comm. on Assessing Crop Yield, Site-Specific farming, Information Systems and Res. Opportunities, Board on Agric. NRC,* Washington: National Academy Press pp.149

IOP Conf. Series: Earth and Environmental Science **275** (2019) 012009 doi:10.1088/1755-1315/275/1/012009

Scanning Probe Microscopy for polymer film characterization in food packaging

F Marinello[1,*], A Pezzuolo[1], D Passeri[2], A La Storia[3], G Mauriello[3]

[1] Department of Land, Environment, Agriculture and Forestry, University of Padova, 35020 Legnaro (PD), Italy.
[2] Department of Basic and Applied Sciences for Engineering, Sapienza University of Rome, Italy
[3] Department of Agricultural Sciences, University of Naples Federico II, Italy

E-mail: francesco.marinello@unipd.it

Abstract. Scanning probe microscopy (SPM) is a branch of microscopy allowing characterization of surfaces at the micro-scale by means of a physical probe (with a size of a few microns) raster scanning the sample. SPMs monitor the interaction between such probe and the surface and, depending on the specific physical principles causing the interaction, they allow generation of a quantitative map of topographic properties: geometrical, optical, electrical, magnetic, etc. This is of the greatest interest, in particular whenever functional surfaces have to be characterized in a quantitative manner. The present paper discusses the different applications of Scanning Probe Microscopy techniques for a thorough characterization of polymer surfaces, of specific interest in particular for the case of food packaging applications.

1. Introduction

Scanning probe microscopy (SPM) refers to a family of techniques which allow imaging of surfaces with nanometric resolution thanks to the implementation of a sharp tip scanned across a sample interface [1]. The specific interaction occurring between such tip and the scanned sample define the microscope variant, which eventually provides the characterization of a given surface property [2]. Typically surface 3D topography is provided by Atomic Force Microscopy (AFM) which measures relies on Van der Waals or other attractive and repulsive forces [3] (Figure 1A). Tip-sample vertical interaction force can be thus converted into a relative vertical elevation, while lateral forces are useful for reconstruction of local friction in the so called Lateral Force Microscopy (LFM) [4] (Figure 1A).
Elastic properties of materials are achieved by Atomic Force Acoustic Microscopy (AFAM): in this case, the sample under investigation is vibrated at ultrasonic frequency, while the cantilever is contacting and scanning the sample surface through the tip [5]. Vibrations perceived by the probe allow dynamic mapping of elastic modulus, shear modulus or Poisson ratio properties (Figure 1C).
Optical information at the nanoscale can be also monitored, through Near-field Scanning Optical Microscopy (NSOM) [6] (Figure 1D). Such technique enables non-diffraction limited imaging and spectroscopy of surfaces thanks to implementation of a small aperture between the light source and the sample surface and the generation of an evanescent field emanated from the aperture in the near-field

IOP Conf. Series: Earth and Environmental Science **275** (2019) 012009 doi:10.1088/1755-1315/275/1/012009

region, at a few tens of nanometers from the aperture. Specific functionalization of the probing tip and of the scanning system allow collection also of other information, such as magnetic or electric properties even at relatively high experimental temperatures [1, 7, 8].

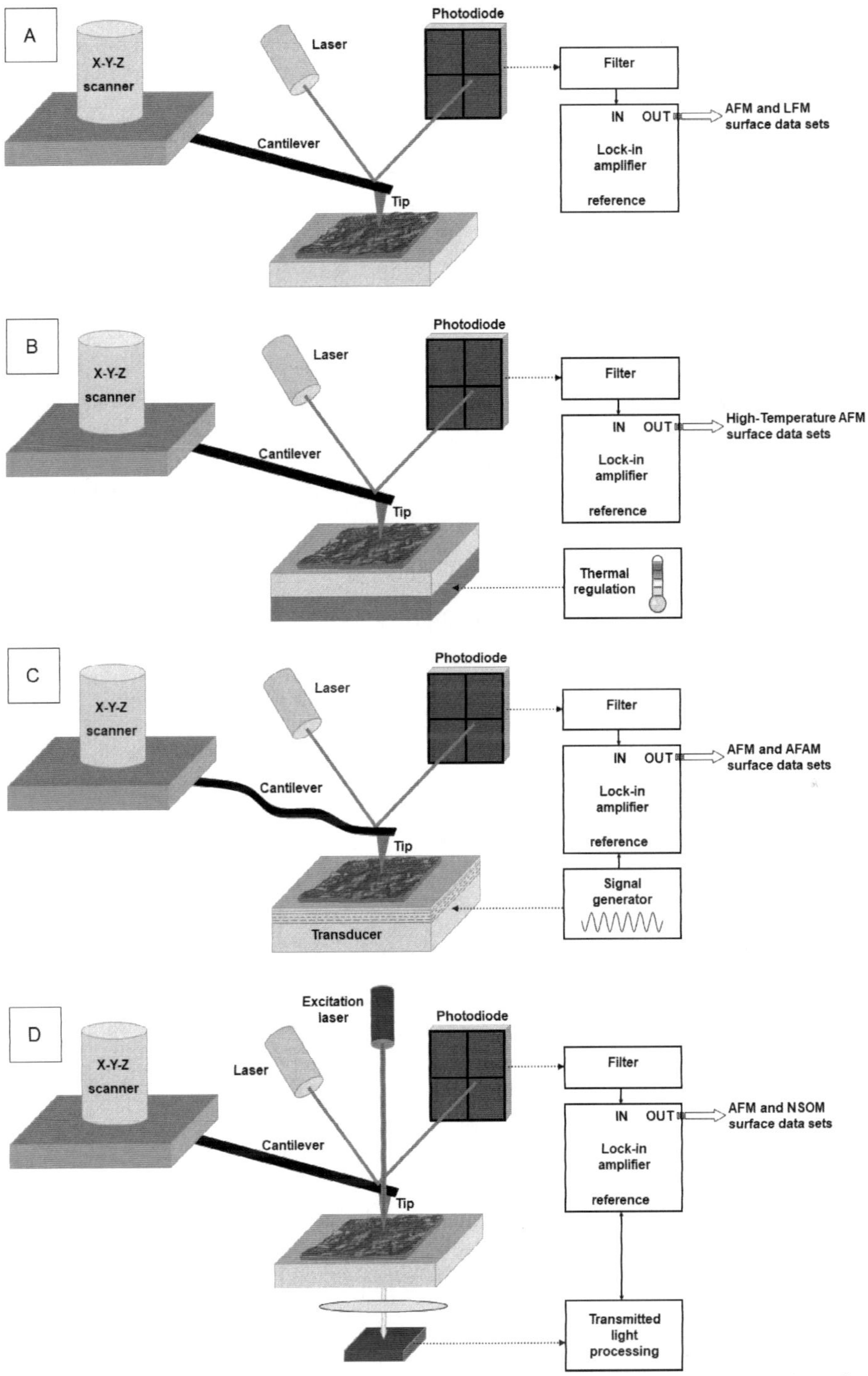

Figure 1. Schematic representation of the four discussed SPM techniques: A) Atomic Force Microscopy and Lateral Force Microscopy; B) High-temperature Atomic Force Microscopy; C) Atomic Force Acoustic microscopy and D) Near-field Scanning Optical Microscopy.

The present paper specifically deals with the applications of Atomic Force Microscopy (AFM), High Temperature Atomic Force Microscopy (HT-AFM), Atomic Force Acoustic Microscopy (AFAM), Lateral Force Microscopy (LFM) and Near-field Scanning Optical Microscopy (NSOM) to polymer film characterization, highlighting the applicative potential for food packaging analyses.

2. Techniques description

2.1. Measuring instruments

SPM is an interesting family of technique allowing characterization both of food packaging materials [10] and of processes occurring on packaging interfaces [11].

Atomic Force Microscopy (AFM) is certainly the most widespread SPM technique, implemented to allow 3D reconstruction at the microscale and to geometrical characterization of micro-features. With regard to food packaging, AFM allows characterization of surface topographies of active packaging which can help preservation of food, such as low roughness, higher permeability or self cleaning structures. Also AFM can help monitoring phenomena occurring at the interface as for instance the evolution of cells surfaces when specific treatments are implemented. This is even more interesting when temperature is modulated, allowing AFM to monitor the packaging film evolution at relatively high temperatures.

Atomic Force Acoustic Microscopy (AFAM) is a dynamic AFM-based technique for fast nanomechanical non-destructive measurements, exploiting the spatial resolution typical of SPMs. AFAM has been widely used not only to analyze the surface elasticity modifications of biological samples, but also for recognizing buried structures in packaging films such as bubbles or nanoparticles which might alter the functional characteristics of the film itself.

Lateral Force Microscopy (LFM) is associated with torsional movements of the probe, which can be associated to frictional properties at the interface. As for AFM, LFM can highlight local dishomogeneities or portions of the surface with higher or lower friction properties which will be related to different levels of interaction between packaging and food.

Aperture Near-field Scanning Optical Microscopy (NSOM) is an interesting technique for high resolution optical imaging of nanostructured surfaces, allowing overcoming the diffraction limit. It can be successfully applied to monitor the distribution of fluorescent nanoparticles, as well as accumulation or propagation of light energy associated with plasmons.

2.2. Measuring tasks

Measuring tasks allowed by the SPM measuring techniques discussed in the previous paragraph have been depicted in Figure 2. The first two lines in the figure are descriptive of geometrical characterizations. In particular, all of the proposed techniques (AFM, HT-AFM, LFM, AFAM and NSOM) allow extraction of information of feature dimensions in terms of:

- areal surface roughness parameters, i.e. indices descriptive of amplitude, functional, volume, hybrid, spatial and miscellaneous parameters, as defined by ISO 25178 [9] (Figure 2A)
- surface waviness and curvatures or shape (Figure 2B)
- distance between isolated or periodical features (Figure 2C)
- height and width of isolated features (Figure 2D and 2E)
- volume of porosity and particles, including nanoparticles layers (Figure 2F and 2G)

All of the previous geometrical characterizations can be repeated at relatively high temperatures (through HT-AFM), in order to monitor and quantify surface evolution over time typically up to 100-150°C (Figure 2H).

The third line in the figure is indicative of other properties which can be measured at the interface, and in particular:

- elastic properties both of bulk materials and thin films (Figure 2I and 2J), as allowed by AFAM techniques

IOP Conf. Series: Earth and Environmental Science **275** (2019) 012009 doi:10.1088/1755-1315/275/1/012009

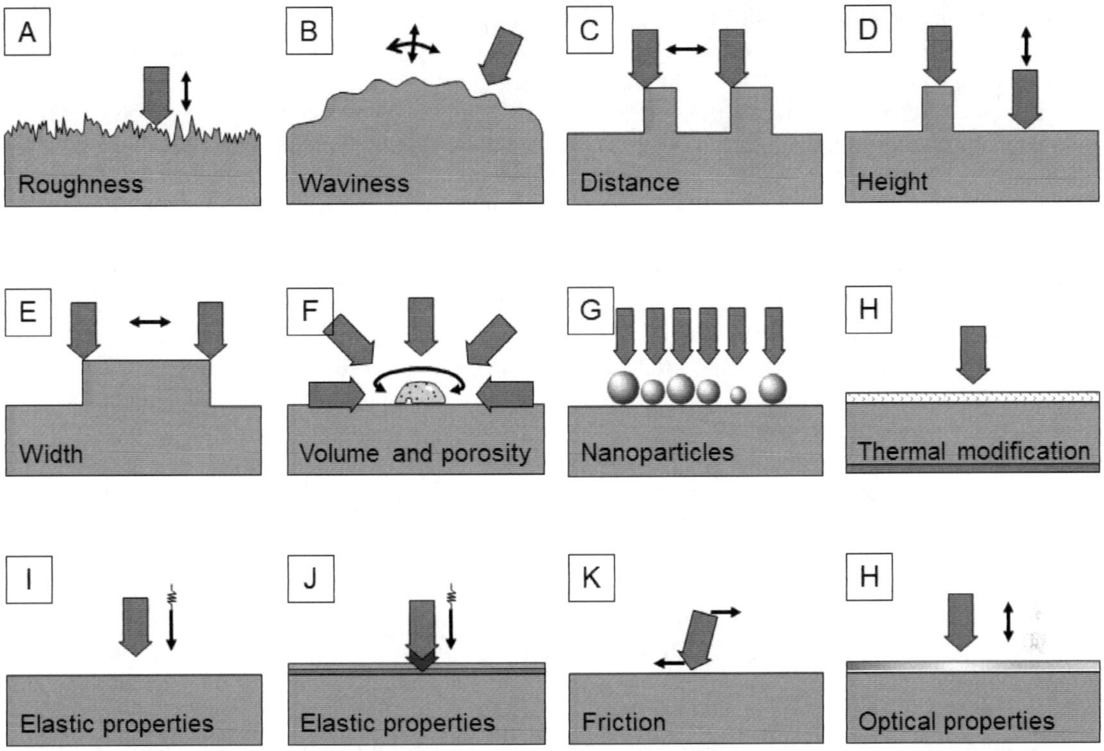

Figure 2. Main tasks to be fulfilled by SPMs techniques. The first two lines refer to extraction of geometry related parameters; the third line refers to other interface properties.

3. Measurement tests

3.1. Tested instruments and polymer films

In order to quantify the performances of different SPM techniques, and their potential in polymer film characterization, six different scanners were involved in the investigation: two Nt-Mdt Ntegra Prima with different configurations (implemented for AFM, HT-AFM, LFM and AFAM), a DME Dualscope DS 95-200 (implemented for AFM, and LFM), a Witec Alpha300 S (for NSOM), an APE Research A100 (for AFM and LFM) and a Veeco Dimension 3100 (for AFM and LFM). Technical specification and drawings have been already reported in a previous work [12].

The different SPM techniques were studied and analysed with different polymer samples which can be representative of different food packaging polymers. In particular the study included the following materials:

- polypropylene with glass fibres (30% FG),
- polymethylmethacrylate and styrene-butadiene-styrene blend (50% PMMA and 50% SBS)
- polycarbonate and acrylonitrile butadiene styrene blend (70% PC and 30% ABS)
- polypropylene and styrene-butadiene-styrene blend (70% PP and 30% SBS)

3.2. Main results

The main results from several experiment on the application of different SPM techniques to different polymer blends are summarized in Table 1.

It can be noticed how Atomic Force Microscopy as a stand alone technique can provide the highest resolution in terms not only of lateral but also of vertical characterization, with minimum measurable features that can be far below 1 nm. The main limitation is ascribable to the fact that only topographical information can be collected, thus related exclusively to geometry and roughness information. In case of relatively high temperature analyses, the same measurements are carried out on heated samples: despite implementation of insulation systems, sample stability and thermal drift can be an issue, causing

an overall reduction of achievable resolution. In this scanning mode, scan time is kept low in order to minimize the influence of thermal related distortions: the effect of high temperatures and high scan rates often result in increased tip wear or contamination phenomena.
Lateral Force Microscopy provide information on local friction properties which might be useful for food-packaging adhesion. Such scan mode relies on lateral bends of the probe: scan operations have the same high performances of AFM measurements, however probe torsions can be affected by many disturbances sources, which reduce the overall signal to noise rate, increasing the difficulty in the interpretation of collected signal.
Exploitation of quantitative Atomic Force Acoustic Microscopy measurements is strictly depending on a number of influencing factors, mainly related to the probe and its interaction with the surface, as for instance tip radius, cantilever stiffness and scan or integration time. Additionally, the tip can undergo severe wear phenomena, mainly due to the strict interaction between the probe and the surface. As in the case of LFM, interpretation of resonance frequency variations and extrapolation of surface elasticity information is typically not easy, and depending on the specifically applied calibration procedures.
Aperture Nearfield Scanning Optical Microscopy relies on large aperture tips, which reduce the minimum achievable resolution and cause large convolution distortions. Also focusing of a laser source inside a probe cavity can rapidly increase the temperature in the very apex of the tip, thus reducing both sample and probe stability.

Table 1. Main SPM performances

Technique	Lateral resolution	Scan time per image (512×512 points)	Main distortions or limiting factors
AFM	<1 nm	<10 min	Only topography information
HT-AFM	<10 nm	<10 min +heating time	Thermal drift, Sample stability, Tip contaminations
LFM	<1 nm	<10 min	Noise, Data interpretation
AFAM	<20 nm	<40 min	Slow scan rate, Tip wear, Data interpretation
NSOM	<50 nm	<30 min +optical alignment	Tip convolution, Thermal drift, Sample and probe stability

4. Conclusions

The present research provides a fast review on most interesting Scanning Probe Microscopy techniques for polymer characterizations. If by one side AFM is the most widespread scanning technique, on the other side different SPM technologies could provide further relevant information, which is useful for the characterization and definition of phenomena at the interface.
Some considerations are proposed on SPMs, based on different measurements carried out with different polymer samples. The comparison in terms of resolution and scanning time and main limiting factors highlight interesting performances which could be implemented in order to profitably characterize polymer surfaces in the food sector.

References

[1] Ma Z, Zhao M, Qu Z, Gao J, Wang F, Shi Y, Qin L and Liu J 2017. Precise nanoscale measurements with scanning probe microscopy (SPM): A review *Journal of Nanoscience and Nanotechnology* **17** 2213-2234.

[2] Passeri D, Rossi M and Vlassak JJ 2013. On the tip calibration for accurate modulus measurement by contact resonance atomic force microscopy. *Ultramicroscopy* **128** 32-41

[3] Passeri D, Biagioni A, Rossi M, Tamburri E and Terranova ML 2013. Characterization of polyaniline-detonation nanodiamond nanocomposite fibers by atomic force microscopy based techniques *European Polymer Journal* **49** 991-998.

IOP Conf. Series: Earth and Environmental Science **275** (2019) 012009 doi:10.1088/1755-1315/275/1/012009

[4] Yang CW, Leung KT, Ding RF, Ko HC, Lu YH, Fang CK and Hwang IS 2018. Lateral Force Microscopy of Interfacial Nanobubbles: Friction Reduction and Novel Frictional Behavior *Scientific Reports* **8** 3125.

[5] Marinello F, Schiavuta P, Vezzù S, Patelli A, Carmignato S and Savio E 2011. Atomic force acoustic microscopy for quantitative nanomechanical characterization. *Wear* **271** 534-538.

[6] Nabha-Barnea S, Maman N, Visoly-Fisher I and Shikler R 2016. Microscopic Investigation of Degradation Processes in a Polyfluorene Blend by Near-Field Scanning Optical Microscopy *Macromolecules* **49** 6439-6444.

[7] Passeri D, Dong C, Reggente M, Angeloni L, Barteri M, Scaramuzzo FA, De Angelis F, Marinelli F, Antonelli F, Rinaldi F, Marianecci C, Carafa M, Sorbo A, Sordi D, Arends IWCE and Rossi M 2014. Magnetic force microscopy: Quantitative issues in biomaterials. BioMatter **4** e29507

[8] Rossi M, Passeri D, Sinibaldi A, Angjellari M, Tamburri E, Sorbo A, Carata E and Dini L 2017 Nanotechnology for Food Packaging and Food Quality Assessment *Advances in Food and Nutrition Research* **82** 149-204.

[9] ISO 25178. Geometrical Product Specifications (GPS) – Surface Texture: Area – Part 2: Terms, Definitions and Surface Texture Parameters, International Organisation for Standardization, Geneva, 2012.

[10] Marinello F, La Storia A, Mauriello G and Passeri D 2018. Atomic Force microscopy techniques to investigate activated food packaging materials. *Trends in Food Science and Technology* **87** 84-93.

[11] La Storia A, Ercolini D, Marinello F and Mauriello G, 2008. Characterization of bacteriocin-coated antimicrobial polyethylene films by atomic force microscopy *Journal of Food Science* **73** T48-T54

[12] Marinello F, Balcon M, Schiavuta P, Carmignato S and Savio E 2011. Thermal drift study on different commercial scanning probe microscopes during the initial warming-up phase. *Measurement Science and Technology* **22** 094016.

IOP Conf. Series: Earth and Environmental Science **275** (2019) 012010 doi:10.1088/1755-1315/275/1/012010

Quality control of hazelnuts by means of NMR measurements

D Di Caro, C Liguori, A Pietrosanto and P Sommella
Department of Industrial Engineering, University of Salerno, Fisciano (SA), Italy
Email: tliguori@unisa.it

Abstract. Hazelnuts are widely used in the confectionary industry for their flavour and taste. In order to guarantee a suitable level of quality, several standards have been defined by international organizations and producing countries. They define the minimum quality requirements of the fruits in terms of dimension, aspect, level of moisture, hidden defects. In this framework, our proposal is related to the set-up of an in-line classification system, based on low field NMR, able to detect the hidden defects of the hazelnuts. The proposed classification procedure is based on the CPMG sequence and the analysis of the transverse relaxation decay. The procedure includes three steps in which different features are detected: (i) moisture content, (ii) kernel development and (iii) healthy detection (presence of mould). Experimental results showed a very good capability to correctly detect the hidden defects, obtaining a sensitivity of 95%, with a specificity (correct detection of the healthy hazelnuts) around the 80%.

1. Main properties and hidden defects of the hazelnuts

Hazelnuts are widely used in the confectionary industry for their flavour and taste. They are characterized by a high nutritional value due to the presence of several components, mainly lipids (about 60% w/w), carbohydrates, proteins, sugar, and dietary fibres [1]. Both the yearly world production and the hazelnuts quality depend on the climate condition. In order to guarantee a suitable level of quality, several standards have been defined by the producing countries and regions. They define the minimum quality requirements of the fruits in terms of dimension, aspect, hidden defects. Other requirements, related to the quality of the hazelnuts and the place of production, can be found, for example, in the European Union certification like the Protected Geographical Indication (PGI). The most recent international standard has been set by the Organization for Economic Co-operation and Development [2]. It defines the quality requirements for the in-shell hazelnuts and hazelnut kernels intended for direct consumption or for food. The minimum requirements are described in the following. The shell must be:

- Intact; only slight superficial damages are allowed.
- Clean; free of any foreign matter.
- Free from blemishes and stains affecting more than 25% of the surface of the shell.
- Well formed.

The kernel must be:

- Free from rancidity.
- Sufficiently developed; kernels should fill at least 50% of the shell cavity.
- Not desiccated.
- Free from blemishes and stains affecting more than 25% of the surface of the shell.
- Well formed.

The whole nut must be:

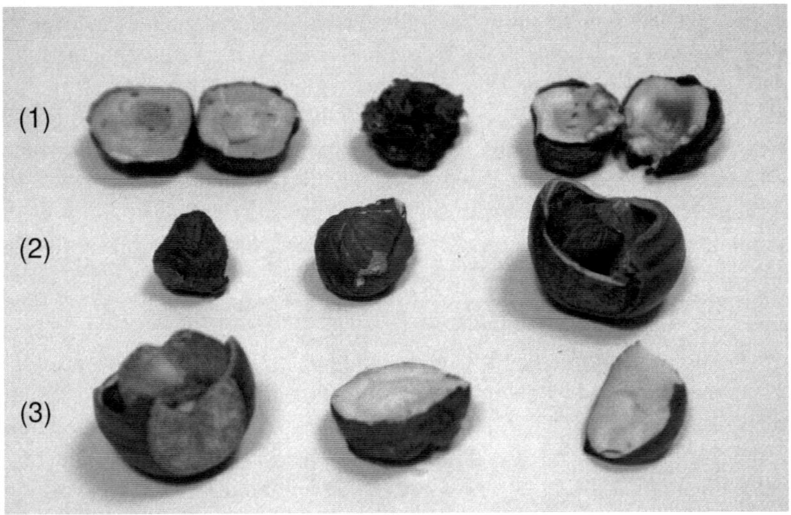

Figure 1. Hazelnut kernels: (1) mould and pest damage; (2) desiccated and not well developed; (3) healthy

- Free from mould filaments.
- Free from living pests.
- Free from damage caused by pests.
- Free of any foreign smell or taste.
- Dried; moisture content not greater than 12% for the whole nut or 7% for the kernel.

The standard also defines the maximum allowed defects related to each requirements providing three quality classes: extra class, class I and class II. Figure 1 shows the aspect of healthy and unhealthy kernels.

2. Techniques to detect hazelnuts quality

The main technique used to detect the hidden defects of the hazelnuts is the visual inspection. A procedure to carry out this kind of test has been defined by the U.S. Food and Drug Administration in the Macroanalytical Procedures Manual (MPM) [3]. The document defines the test procedures for several food products, like fruits, vegetables, grain, dairy, cheese and seafood. It defines:

- Sample preparation.
- Sequential sampling plans.
- Visual and organoleptic examination.
- Classification of reject nuts.
- Report.

This method, being based on visual and organoleptic analysis, requires time and skilled personnel for a correct nut classification. Other techniques have been investigated in order to achieve automatic systems to check the quality parameters of the hazelnuts. The dimension of the shell and the blank nuts are detected by means of machines based on mechanical techniques. In particular, calibrator machines able to sort the whole hazelnuts on the basis of their diameter are available making use of sieves with different diameters. Air compressed-based machines can detect and separate some kinds of foreign body, like leaves or pieces of wood, from the whole hazelnuts. They can also detect blank nuts, even if they are not accurate. These kinds of equipment are useful for a pre-treatment of the product but they do not represent quality control systems. A different method to detect empty hazelnuts, based on the analysis of the acoustic signal generated by the impact of the nut on a steel plate, has been proposed [4]. It exploits both time-domain and frequency-domain analysis of the signal and can be used for a real-time detection. Using the same principle, a sorting method of cracked, hollow and regular shell has been developed for hazelnuts [5] and pistachio nuts [6], exploiting Frequency Domain (FD) signal processing techniques and Artificial Neural Network (ANNs).

Several studies have been conducted related to the moisture content determination. This is an important quality parameter because a high level of moisture causes fungal diseases and mould formation, and then postharvest losses. Based on the weather condition around the harvest period, the moisture content of the hazelnuts can be more than 30%. They are dried using industrial dryer or, often, by means of sun drying techniques. After this process, farmers need instruments to verify that the moisture content is lower than the limits imposed by the quality standards. The most common method to evaluate the moisture content in food is based on oven drying in which the sample is heated under specified conditions and the amount of moisture is determined calculating the loss of weight:

$$\%Moisture = \frac{weight\ wet\ sample - weight\ dry\ sample}{weight\ wet\ sample} \times 100 \qquad (1)$$

Other methods, related in particular to seeds and nuts, have been proposed. RF impedance measurement has been employed to determine the moisture in single grain and peanut kernels placed in a small parallel plate capacitor [7]. A similar method has been applied to determine the amount of moisture in in-shell hazelnuts placed in a probe with two vertical parallel-plate electrodes able to host a sample of 250 g of nuts [8]. More complex analytical techniques have been used to evaluate some quality characteristics of the hazelnuts. For example, a study for the identification of chemical markers of hazelnut roasting, based on Headspace Solid Phase Microextraction (HS-SPME) coupled with Gas Chromatography – Mass Spectrometric (GC-MS) detection has been proposed [9]. Near Infrared (NIR) Spectroscopy has also been used to detect flawed kernels and to estimate lipid oxidation [10]. These kinds of techniques allow extracting complex features in a non-destructive way, but are suitable for laboratory testing and cannot used for industrial applications.

2.1 Using NMR in food quality control

NMR has been mainly used, in the past, in chemical analysis and medical diagnosis. The applications in food quality control have been developed progressively, thanks to the technical improvement in the instrumentation, the signal processing and especially the development of permanent magnets that allowed a significant cost reduction of the NMR equipment.

The NMR systems can be classified, on the basis of the technique used to carry out the analysis, in three main categories:

- NMR Spectrometry.
- NMR Relaxometry.
- MRI (Magnetic Resonance Imaging).

NMR Spectrometry, also named High Field NMR or Fourier Transform NMR (FT-NMR), makes use of superconducting magnets with a static magnetic field usually higher than 1 T. It is based on the frequency domain analysis of the NMR signals and it allows obtaining quantitative information related to the chemical composition of the samples under test.

Relaxometry mainly differs from the spectrometry for the use of permanent magnets that makes it the cheapest NMR technique and, for this reason, the most interesting to investigate industrial applications. It is also named Low Field NMR or Time Domain NMR (TD-NMR) because it employs magnets with a static magnetic field that is usually lower than 1 T and the signal analysis, differently from the spectrometry, is carried out in the time domain.

All of the above techniques have been investigated in food analysis. An overview of the food applications of the liquid state FT-NMR can be found in [11]. It presents a review of the NMR methodologies in food quality control, analysing several aspects as the sample preparation, which is an important step to obtain reproducible spectra, the spectral analysis and the statistical analysis. It also describes some applications to several products like alcoholic beverage, fruits and vegetables, milk and dairy products. An example of food authentication is described in [12], in which a bench-top spectrometer has been used for the analysis of olive oil with adulteration of hazelnut oil. A study on the compositional analysis of wine in a full intact bottle has been proposed in [13]. In the paper, the capability to detect acetic acid spoilage has been investigated using NMR spectrometry system. An overview of the potential applications of NMR and MRI for compositional and structural analysis, inspection of microbiological, physical and chemical quality, food authentication, on-line monitoring

of food processing is presented in [14]. A review focused on the development of TD-NMR and MRI based sensors for different food applications, can be found in [15].
Several studies have been carried out on methods and applications of Low Field NMR. A method for Solid Fat Content (SFC) and simultaneous oil and moisture determination has been described in [16]. Applications on quality control of different food products have been proposed: the evaluation of quality of oranges during storage [17]. Analysis of water dynamic states and age-related changes in mozzarella cheese [18]. The interest towards the industrial applications of the TD-NMR is constantly increasing. A study of the perspective of the use of the TD-NMR as in-line industrial sensor can be found in [19]. Among the proposed applications in in-line monitoring, Low Field NMR has been employed for the evaluation of the internal browning of whole apples [20]. Moreover, an automated system for oil and water content determination in corn seeds has been proposed in [21].

3. The adopted TD-NMR measurement system

The main applications of the TD-NMR in food quality control are related to the evaluation of the presence of water, moisture, oil, fat, that are well detected by this technique. The quality of the hazelnuts is strictly related to the presence of moisture and fatty acids and then, in this work, Low Field NMR has been taken into account as a suitable technique for the analysis of the hazelnuts quality [22]. The TD-NMR instrument used for the experimental work is composed by three main components (Fig. 2): (i) NMR console; (ii) Magnet and probe; (iii) Personal computer.
The NMR console is manufactured by Spincore Technologies Inc. It includes all the electronics to perform the generation and transmission of the excitation pulse sequences as well as the acquisition of the RF signals from the probe. It allows employing a resonance frequency up to 37.5 MHz.
These components have been designed for the specific application considering that the probe, placed in the magnet, has to be able to host a whole hazelnut. To this aim, the probe has a cylindrical coil characterized by height and diameter both equal to 25 mm.
Considering the value of the static magnetic field, the resonance frequency of the system is 21.8 MHz. The field homogeneity of the magnet is 10^{-4} $\Delta B_0/B_0$. This value is only guaranteed in an area with 25 mm diameter at the centre of the magnet. For this reason, the coil of the probe must be placed at the centre of the magnet in order to work with the highest field homogeneity. The NMR console is connected to a personal computer and a Labview user interface has been implemented to send the setting commands to the NMR console and to carry out the signal processing and the classification algorithm on the received signals.
Permanent magnets have a coefficient of temperature in the order of -1000 ppm/°C, so the static magnetic field, and then the resonance frequency, changes when the temperature changes. In order to limit the resonance frequency variation due to the temperature variation, the magnet has been equipped with a temperature controller and insulated by means of sheets of extruded polystyrene.
In this way, when the operating temperature is kept at 23.0 ± 0.1 °C the resonance frequency variation is lower than 5 kHz around the nominal value [23].

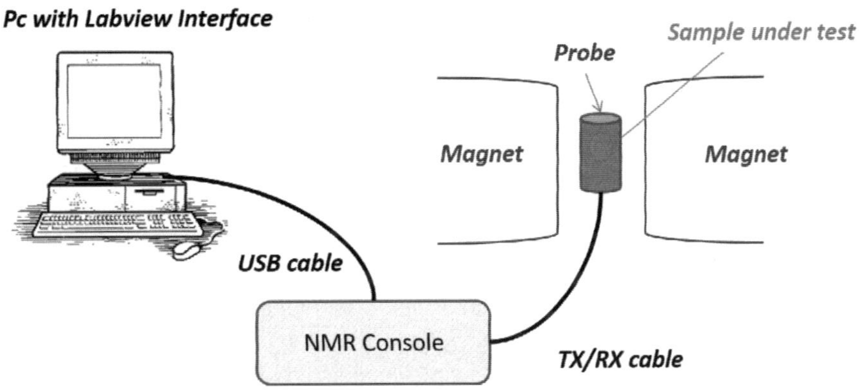

***Figure* 2** The Hardware of the proposed system

IOP Conf. Series: Earth and Environmental Science **275** (2019) 012010 doi:10.1088/1755-1315/275/1/012010

4. The classification algorithm

The proposed classification algorithm for the in-shell hazelnuts is composed by three steps (Fig.3)

1. Evaluation of the moisture content;
2. Evaluation of the kernel development;
3. Evaluation of the presence of mould.

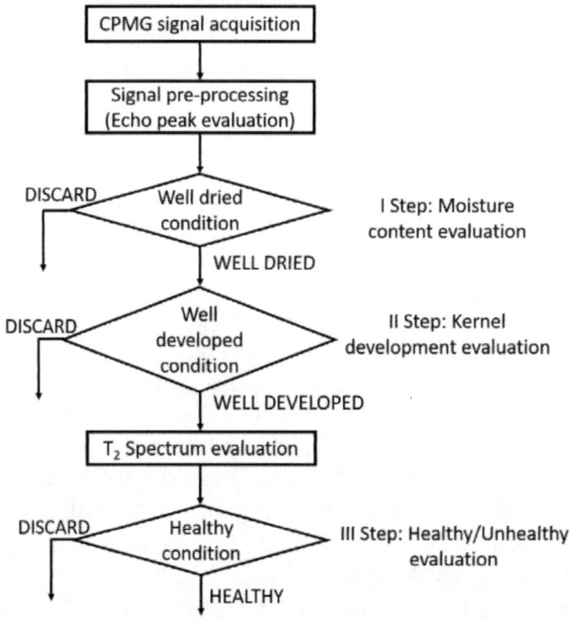

Figure 3 Block diagram of the classification algorithm

In the first step, the hazelnuts with a moisture content greater than the limit imposed by the quality standards are discarded; the correctly dried hazelnuts are then evaluated on the basis of the kernel development. In this step, the following defects are detected and discarded:

- Empty hazelnuts;
- Not well developed kernels;
- Desiccated kernels.

Finally, in the third step, the presence of mould is evaluated and the healthy hazelnuts are selected.
The three steps are carried out on a single signal, acquired by means of the CPMG sequence, so for each sample, only one CPMG sequence is required [24], [25].
The tests were conducted on hazelnuts supplied by "Consorzio di Tutela Nocciola di Giffoni I.G.P.".
For each sample of hazelnut with shell, the following procedure was used to perform the test:

- Insertion of the hazelnut with shell inside the probe;
- Acquisition of the signal by means of CPMG sequence;
- Signal processing on the data acquired by the instrument and classification of the hazelnut by means of classification algorithm; extraction of the hazelnut from the probe;
- Breakage of the shell and classification of the hazelnut as healthy or unhealthy by means of vision inspection;
- Verification of the classification algorithm comparing the results with the vision inspection method.

The tests were carried out on 300 samples of hazelnuts, 220 dried samples and 80 fresh samples.

4.1. Moisture content evaluation

The moisture content evaluation is made analyzing the relationship between the amplitude of the FID signal (A_{FID}) and the amplitude of the 2nd echo, A_{echo}, of the CPMG sequence (Fig. 4) [26].

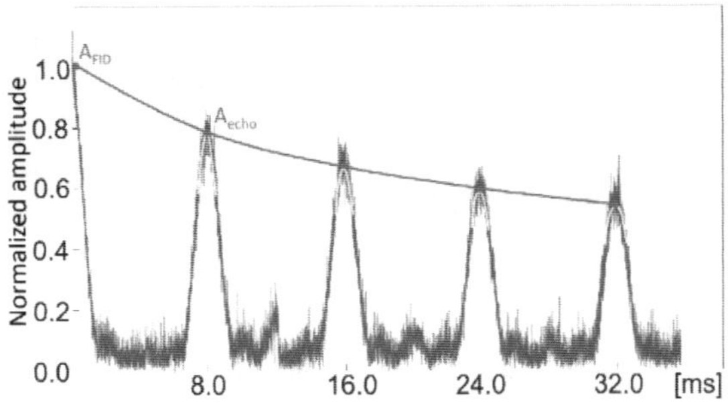

Figure 4 The transverse relaxation decay (red) as envelope of the even echo peaks (yellow) calculated on the received signal (blue)

According to the threshold line defined in (eq. 2), the well-dried hazelnuts have to satisfy the following condition:

$$A_{echo} > (m - 3\sigma_m)A_{FID} + (b - 3\sigma_b) \qquad (2)$$

The results obtained on 220 dried hazelnuts and 80 fresh hazelnuts are shown in Figs 5.

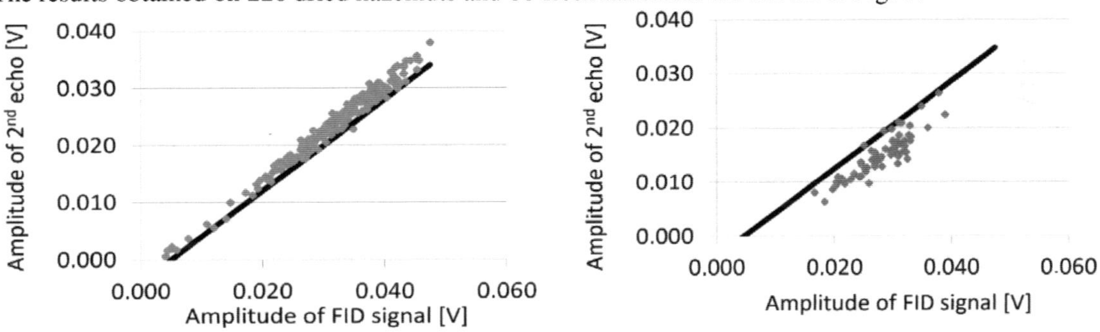

Figure 5 (a) Comparison among the dry threshold line (in black) and the amplitudes of the FID signal and the 2nd echo peak for 220 dried hazelnuts samples (in blue); (b) Comparison among the dry threshold line (in black) and the amplitudes of the FID signal and the 2nd echo peak for 80 fresh hazelnuts samples (in red).

As can be seen from the figure one in 220 samples of dried hazelnuts has been mis-detected, while the 80 fresh hazelnuts have been correctly classified.

4.2. Kernel development evaluation

The analysis on the kernel development on the 219 hazelnuts correctly classified as dried in the previous step is described. It is made exploiting the threshold on the amplitude of the FID (Fig. 6). Three category of defects are considered: blank, insufficiently developed and desiccated have been included in the not well developed kernel. The well-formed hazelnuts have to satisfy the condition:

$$A_{FID} > threshold \qquad (3)$$

In Fig. 6, the results related to the 219 samples selected from the previous step are shown.

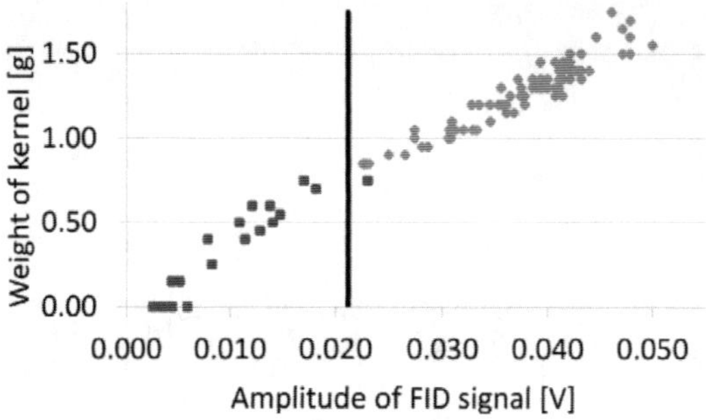

Figure 6 Relation between the amplitude of the FID signal and weight of the kernel for 219 hazelnuts samples: fully developed kernels (in blue) and not well developed kernels (in red)

As can be seen, all the well-developed kernels have been detected and only one in 25 not well developed kernels has been not correctly classified.

4.3. Mould presence evaluation

The analysis of the mould presence is made on the dried and well-developed hazelnuts. Analysing the T_2 spectrum, calculated considering a model of the transverse relaxation curve with three exponential decay [22]:

$$y(t) = \sum_{j=1}^{3} x_j e^{-\frac{t}{T_{2j}}} \quad (4)$$

The healthy hazelnuts exhibited a greater value of the second time constant than the unhealthy ones, although the two categories are not well separated (Fig. 7).

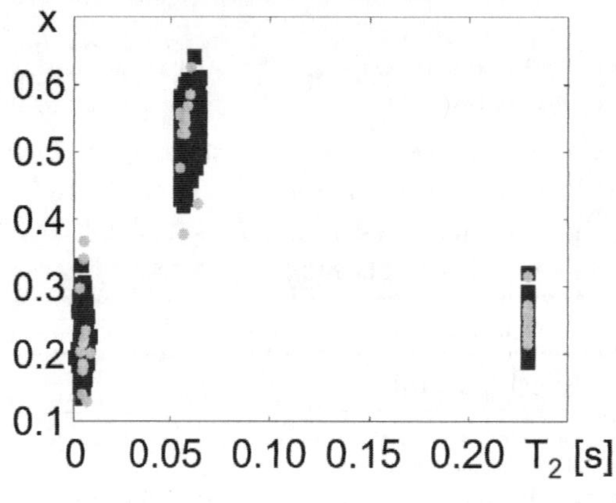

Figure 7 T_2 spectrum for healthy (blue) and unhealthy (green) hazelnuts

Starting from this consideration, a healthy condition has been defined to detect the healthy hazelnuts:

$$T_{2,2} > T_{2,2}^{THR} \quad (5)$$

The threshold has been determined considering an approach based on ROC curves [27]. The analysis has been carried out considering two classes:

- Unhealthy hazelnuts: positive test;
- Healthy hazelnuts: negative test.

The parameters used to evaluate the classification errors are described in Table I.

Table I *Classification parameters and indexes*

Classification parameter	Description	Index	Description
True positive, TP	Unhealthy sample correctly classified	True Posite rate TPR	$TPR = \frac{TP}{TP + FN}$
False positive, FP	Healthy sample classified as unhealthy	False Positive Rate FPR	$FPR = \frac{FP}{FP + TN}$
True negative, TN	Healthy sample correctly classified	True negative rate TNR	$TNR = \frac{TN}{FP + TN}$
False negative, FN	Unhealthy sample classified as healthy		

The true positive rate (TPR), also named sensitivity or recall, represents the capability of the system to detect the unhealthy samples. The false positive rate (FPR) is a measure of the false alarms, namely the healthy samples wrongly classified as unhealthy. The true negative rate (TNR) is the specificity, which represents the capability of the system to detect the healthy hazelnuts.
The ROC curve is a two-dimensional graph in which the y-axis is represented by the TPR and x-axis is represented by the FPR; each point of the ROC curve is related to a value of the threshold to set, and it is obtained calculating the TPR and FPR using that value of the threshold.
Figure 8 shows the ROC curve related to the classification of the hazelnuts obtained varying the threshold THR

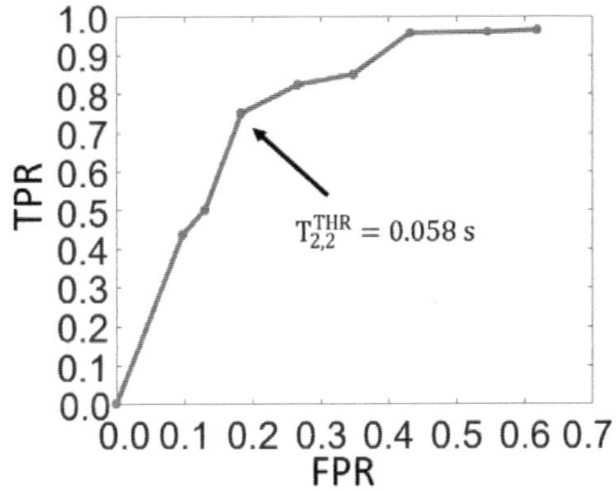

Figure 8 ROC curve related to the threshold $T_{2,2}^{THR}$

Analysing the curve, it can be observed a linear behaviour until a TPR value around 0.80 and FPR around 0.20, then the slope drastically decreases. It means that from that point, the FPR increases faster than the TPR, causing a performance degradation. The threshold $T_{2,2}^{THR} = 0.058$ s has been selected.
Considering the global performance of classification algorithm, a sensitivity (TPR) of 95.0% and a specificity (TNR) of 80.0% has been obtain. The classification algorithm exhibited a high sensitivity, so a high capability to detect the unhealthy hazelnuts.

5. Conclusions

In this work, a method for the quality detection of the in-shell hazelnuts has been proposed. It is based on the low field NMR, which is a non-destructive technique employed in the quality control of food, especially in offline laboratory applications. This research project focuses on an in-line industrial application of the NMR system that has to be able to detect the hidden defects of the hazelnuts in order to help the farmers and the producers to meet the quality requirements imposed by the international standards and rules. A permanent magnet and probe specifically designed for this application have been used in order to be able to host a whole nut. The proposed classification algorithm is based on the analysis of the transverse relaxation decay, which is obtained with the CPMG sequence. In particular, three different steps of the algorithm have been defined. In each step a single quality property of the hazelnuts is analysed: the moisture content, which is responsible of the mould development, is

evaluated analysing the relation between the maximum amplitude of the CPMG signal and the second echo peak. The kernel development, that allows detecting the empty, not well developed and desiccated nuts, is evaluated considering the maximum amplitude of the acquired signal. Finally, the mould presence is evaluated analysing the T_2 spectrum of the CPMG signal, which consists of the amplitudes and the time constants related to the multi-exponential decomposition of the transverse relaxation decay. The first two features can be detected in a very fast way, only 12 ms are needed to obtain the results, because the information are contained in the first part of the acquired signal; for the third feature, the total acquisition time is needed, which is around 245 ms.
The classification algorithm showed a high sensitivity (95%) with a specificity around the 80%, so it is capable detecting the unhealthy hazelnuts but with a relatively high number of discarded healthy hazelnuts. The loss of specificity is mainly due to the third step of the algorithm, while in first two steps this value is similar to the sensitivity. This does not represent a limit because the defects related to the kernel development are well detected and they outnumber other kinds of defects. Moreover, the moisture content is also well detected by the classification algorithm, and this allows discarding the hazelnuts that contain a high level of moisture, that causes the mould development during the storage.
Finally, the study has been carried out on the hazelnuts, but it can be easily extended to other kinds of fruit in shell, as well as the measurement algorithms can be applied in other kinds of NMR applications.

References

[1] Memoli A, Albanese D, Esti M, Lombardelli C, Crescitelli A, Di Matteo M, Benucci I. Effect of bug damage and mold contamination on fatty acids and sterols of hazelnut oil. *European Food Research and Technology*. 2017 Apr 1;**243**(4):651-8.
[2] OECD (2011), International Standards for Fruit and Vegetables: Inshell hazelnuts and hazelnut kernels, International Standards for Fruit and Vegetables, OECD Publishing. http://dx.doi.org/10.1787/9789264166721-en-fr
[3] FDA Technical Bulletin Number 5 Macroanalytical Procedures Manual 1984; Electronic Version 1998
[4] Onaran I, Dulek B, Pearson TC, Yardimci Y, Cetin AE. Detection of empty hazelnuts from fully developed nuts by impact acoustics. InSignal Processing Conference, 2005 13th European 2005 Sep 4 (pp. 1-4). IEEE.
[5] Kalkan H, Çetisli B. Online feature selection and classification. InAcoustics, Speech and Signal Processing (ICASSP), 2011 IEEE International Conference on 2011 May 22 (pp. 2124-2127). IEEE.
[6] Mahdavi-Jafari S, Salehinejad H, Talebi S. A pistachio nuts classification technique: An ann based signal processing scheme. InComputational Intelligence for Modelling Control & Automation, 2008 International Conference on 2008 Dec 10 (pp. 447-451). IEEE.
[7] Nelson SO, Kandala CV, Lawrence KC. Single-kernel moisture determination in grain and nuts by RF impedance measurements. InInstrumentation and Measurement Technology Conference, 1992. IMTC'92., 9th IEEE 1992 May 12 (pp. 422-424). IEEE.
[8] Solar M, Solar A. Non-destructive determination of moisture content in hazelnut (Corylus avellana L.). *Computers and Electronics in Agriculture*. 2016 Feb 1;**121**:320-30.
[9] Nicolotti L, Cordero C, Bicchi C, Rubiolo P, Sgorbini B, Liberto E. Volatile profiling of high quality hazelnuts (Corylus avellana L.): chemical indices of roasting. *Food chemistry*. 2013 Jun 1;**138**(2-3):1723-33.
[10] Pannico A, Schouten RE, Basile B, Romano R, Woltering EJ, Cirillo C. Non-destructive detection of flawed hazelnut kernels and lipid oxidation assessment using NIR spectroscopy. *Journal of Food Engineering*. 2015 Sep 1;**160**:42-8.
[11] Mannina L, Sobolev AP, Viel S. Liquid state 1H high field NMR in food analysis. *Progress in Nuclear Magnetic Resonance Spectroscopy*. 2012 Oct;**66**:1.
[12] Parker T, Limer E, Watson AD, Defernez M, Williamson D, Kemsley EK. 60 MHz 1H NMR spectroscopy for the analysis of edible oils. TrAC *Trends in Analytical Chemistry*. 2014 May 1;**57**:147-58.

[13] Weekley AJ, Bruins P, Sisto M, Augustine MP. Using NMR to study full intact wine bottles. *Journal of Magnetic Resonance*. 2003 Mar 1;**161**(1):91-8.

[14] Marcone MF, Wang S, Albabish W, Nie S, Somnarain D, Hill A. Diverse food-based applications of nuclear magnetic resonance (NMR) technology. *Food Research International*. 2013 May 1;**51**(2):729-47.

[15] Kirtil E, Cikrikci S, McCarthy MJ, Oztop MH. Recent advances in time domain NMR & MRI sensors and their food applications. *Current Opinion in Food Science*. 2017 Oct 1;**17**:9-15.

[16] Todt H, Guthausen G, Burk W, Schmalbein D, Kamlowski A. Water/moisture and fat analysis by time-domain NMR. *Food Chemistry*. 2006 Jun 1;**96**(3):436-40.

[17] Zhang J, Liu Y, Wang N, Ruan R. NMR Technique Application in evaluating the Quality of navel orange during storage. *Procedia Engineering*. 2012 Jan 1;**37**:234-9.

[18] Gianferri R, D'Aiuto V, Curini R, Delfini M, Brosio E. Proton NMR transverse relaxation measurements to study water dynamic states and age-related changes in Mozzarella di Bufala Campana cheese. *Food chemistry*. 2007 Jan 1;**105**(2):720-6.

[19] Colnago LA, Andrade FD, Souza AA, Azeredo RB, Lima AA, Cerioni LM, Osán TM, Pusiol DJ. Why is inline NMR rarely used as industrial sensor? Challenges and opportunities. *Chemical Engineering & Technology*. 2014 Feb;**37**(2):191-203.

[20] Chayaprasert W, Stroshine R. Rapid sensing of internal browning in whole apples using a low-cost, low-field proton magnetic resonance sensor. *Postharvest biology and technology*. 2005 Jun 1;**36**(3):291-301.

[21] Wang H, Liu J, Xu X, Huang Q, Chen S, Yang P, Chen S, Song Y. Fully-automated high-throughput NMR system for screening of haploid kernels of maize (Corn) by measurement of oil content. *PloS one*. 2016 Jul 25;**11**(7):e0159444.

[22] Di Caro D (2018), NMR measurements for hazelnuts classification (Doctoral dissertation) http://dx.doi.org/10.14273/unisa-1394.

[23] Di Caro D, Liguori C, Pietrosanto A, Sommella P. Using a SVD-based algorithm for T 2 spectrum calculation in TD-NMR application to detect hidden defects in hazelnuts. InInstrumentation and Measurement Technology Conference (I2MTC), 2017 IEEE International 2017 May 22 (pp. 1-6). IEEE.

[24] Di Caro D, Liguori C, Pietrosanto A, Sommella P. Investigation of the NMR techniques to detect hidden defects in hazelnuts. InInstrumentation and Measurement Technology Conference Proceedings (I2MTC), 2016 IEEE International 2016 May 23 (pp. 1-5). IEEE.

[25] Di Caro D, Liguori C, Pietrosanto A, Sommella P. Hazelnut Oil Classification by NMR Techniques. *IEEE Transactions on Instrumentation and Measurement*. 2017 May;**66**(5):928-34.

[26] Di Caro D, Liguori C, Pietrosanto A, Sommella P. Advanced TD-NMR based estimation of the inshell hazelnuts quality parameters. In2018 IEEE International Instrumentation and Measurement Technology Conference (I2MTC) 2018 May 14 (pp. 1-6). IEEE.

[27] Fawcett T. An introduction to ROC analysis. *Pattern recognition letters*. 2006 Jun 30;**27**(8):861-74.

IOP Conf. Series: Earth and Environmental Science **275** (2019) 012011 doi:10.1088/1755-1315/275/1/012011

Application of ISO 25178 standard for multiscale 3D parametric assessment of surface topographies

F Marinello[1,*], A Pezzuolo[1]

[1] Department of Land, Environment, Agriculture and Forestry, University of Padova, 35020 Legnaro (PD), Italy.

E-mail: francesco.marinello@unipd.it

Abstract. The objective of the present work is to discuss the potential of areal surface texture parameters as introduced and discussed by ISO standards 25178, as a widely recognized reference framework of indices and procedures, which can help and accelerate understanding of functional information. Such indices have been developed specifically for the micro-scale, however they can be successfully implemented also in the case of larger scales. Parameters extraction takes place in three main steps, independently from the scale: calibration, filtering and parameter extraction. The possibility of using the same approach and the same roughness parameters at different scales helps very much not only the post processing of surfaces data sets but also their interpretation, putting the basis for multiscale models.

1. Introduction

In the last couple of decades, the growing availability of sensors and instruments installed in laboratories or in the field (at ground level or on board of drones, airplanes or satellites) has greatly increased the amount of available data. In all of the cases, surface roughness is often addressed as a key parameter, useful to characterize processes (at the micro-scale), determine shapes (at the meso-scale) and identify landforms (at the macro-scale).

By way of example, roughness parameters have been used in order to delimit landslides [1], identify road networks [2], characterize animal body surface features [3, 4], analyse small devices functionality [5] or study nanostructured food packaging [6].

A number of different instruments has been developed in order to allow quantitative characterization of surface topographies at different scales. Such techniques are typically relying on optical non contact methods, which guarantee minimization of scanning times and of interaction forces with the measurand, and include LiDAR, diffraction, fringe projection, triangulation, interferometry, confocal etc [7-9]. In all of the cases, instrumentation collects local elevation on a vertical z-axis (as digital elevation models, DEM) producing an array of points regularly organized in the horizontal x-y plane. Despite a variety of terminology has been applied (ruggedness, rugosity, microrelief or microtopography), surface roughness is the generic term which identifies and characterizes topographies [10].

Data are very often structured in the form of maps with vertical elevation or some different functional information reproduced as pixels with constant resolution and spacing. Such maps are undoubtedly increasing the level of knowledge of different processes in agricultural or forestry fields, but, on the

other hand, might cause growing difficulties and efforts, especially during data post processing. For this reason, parametrization is a fundamental step, which can help exploiting the content of information brought by revealed data. Through parametrization, big amounts of data can be summarized in a few numbers which accelerate understandability and in general decision making processes.
The main objective of the present work is to discuss the potential of areal surface texture parameters as introduced and discussed by ISO standards 25178 [11], as a widely recognized reference framework of indices and procedures which can help and accelerate understanding of functional information. ISO 25178 has been developed by the technical committee TC 213 (working group WG 16) for all areal surface topography measuring instruments regardless of their design or operation. Such indices have been developed specifically for the micro-scale, however they can be successfully implemented also in the case of larger scales. The possibility of using the same approach and the same roughness parameters at different scales helps very much not only the post processing of surfaces data sets but also their interpretation, putting the basis for multiscale models.

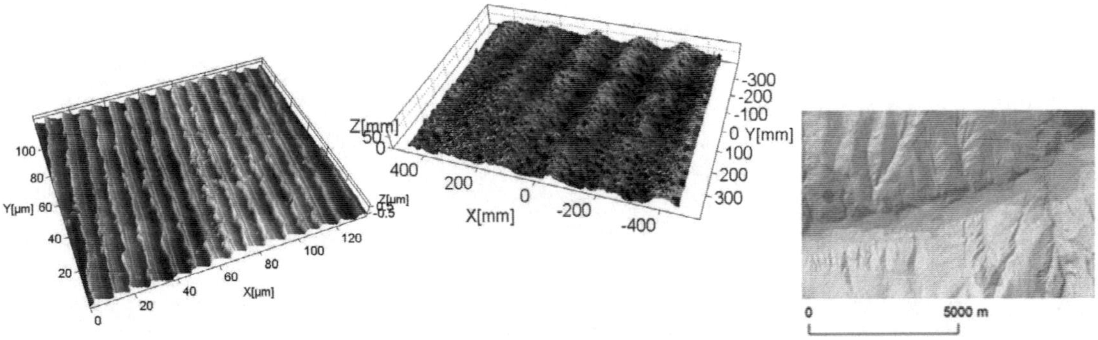

Figure 1. Three examples of topographies at three different scales: a micromilled surface in a 130×110µm range, a harrowed soil portion in a 900×700mm and a valley topography in a 9000×6000m.

2. Approach description

2.1. Data processing procedure
The extraction of quantitative parameters can take place according to the following three main steps, which are independent from the scale, and can be in general applied for post-processing of surface topographies:
1) calibration, to eliminate systematic distortions and give metrological traceability to measured data
2) three-dimensional filtering, as a needed set of operations allowing elimination of distortions (caused by the measurement process) and exploiting repeatability in the subsequent extraction of indices
3) extraction of different roughness parameters, including: height parameters (root mean square roughness, kurtosis, skewness,...), function related parameters (material ratio, volume,...), hybrid parameters (interfacial area ratio, root mean square gradient,...), spatial parameters (autocorrelation functions, texture direction,...).
A graphical representation of the three main operations following data acquisition is proposed in the following flow chart (Figure 2).

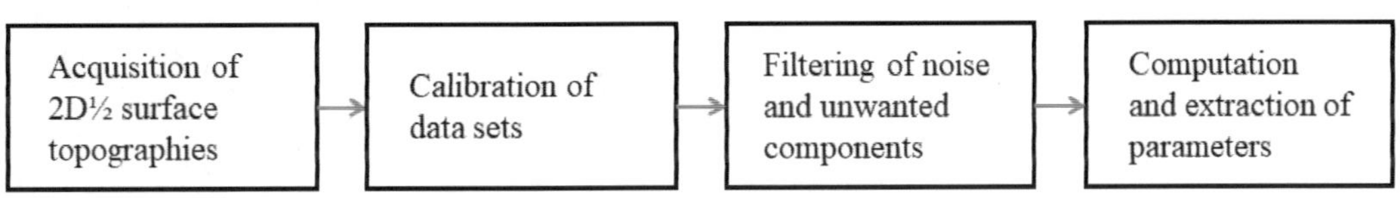

Figure 2. Flow chart of the proposed data processing approach

2.2. Calibration

In this paragraph a complete geometrical model for calibration of data sets is presented, adapted from [12]. In order to better clarify the model, basic assumptions are here briefly given. Let $i = 1, \dots, m$, and $j = 1, \dots, n$ be indices of the scanned pixels in a matricial representation, and x_i', y_j' and z' the corresponding lateral and vertical positions in a reference *x-y-z* space.
The apostrophe ' indicates that the measuring coordinate axes of the optical instrument are taken as reference coordinate system. Also it is assumed that the transfer function from the uncorrected z-coordinate $z'(x',y')$ to the actual coordinate $z(x',y')$ is single valued: as a consequence each measured position (x',y') corresponds to exactly one corrected value.
Under such condition, the following model based on the McLaurin series, evaluated in $(0,0,0)$ can be introduced to describe and calibrate surface topography data sets,:

$$x(x',y',z') = \sum_{n_{x'}=0}^{\infty}\sum_{n_{y'}=0}^{\infty}\sum_{n_{z'}=0}^{\infty} \frac{\partial^{n_{x'}} x}{\partial x'^{n_{x'}}} \frac{\partial^{n_{y'}} x}{\partial y'^{n_{y'}}} \frac{\partial^{n_{z'}} x}{\partial z'^{n_{z'}}} \frac{x'^{n_{x'}} y'^{n_{y'}} z'^{n_{z'}}}{n_{x'}!\,n_{y'}!\,n_{z'}!}$$

Approximating the series to the third order, and introducing the Peano residual term σ to include higher order infinitesimals, the following expression can be written:

$$x(x',y',z') = \frac{\partial x}{\partial x'}x' + \frac{\partial x}{\partial y'}y' + \frac{\partial x}{\partial z'}z' + \frac{\partial^2 x}{\partial x'^2}\frac{x'^2}{2} + \frac{\partial^2 x}{\partial y'^2}\frac{y'^2}{2} + \frac{\partial^2 x}{\partial z'^2}\frac{z'^2}{2} + \frac{\partial^2 x}{\partial x'\partial y'}\frac{x'y'}{2} + \frac{\partial^2 x}{\partial x'\partial z'}\frac{x'z'}{2} + \frac{\partial^2 x}{\partial y'\partial z'}\frac{y'z'}{2} + \frac{\partial^3 x}{\partial x'^3}\frac{x'^3}{6} + \frac{\partial^3 x}{\partial y'^3}\frac{y'^3}{6} + \frac{\partial^3 x}{\partial z'^3}\frac{z'^3}{6} +$$
$$\frac{\partial^3 x}{\partial x'^2\partial y'}\frac{x'^2y'}{6} + \frac{\partial^3 x}{\partial x'^2\partial z'}\frac{x'^2z'}{6} + \frac{\partial^3 x}{\partial y'^2\partial z'}\frac{y'^2z'}{6} + \frac{\partial^3 x}{\partial x'\partial y'^2}\frac{x'y'^2}{6} + \frac{\partial^3 x}{\partial x'\partial z'^2}\frac{x'z'^2}{6} + \frac{\partial^3 x}{\partial y'\partial z'^2}\frac{y'z'^2}{6} + \frac{\partial^3 x}{\partial x'\partial y'\partial z'}\frac{x'y'z'}{6} + \sigma\left((x,y,z)^3\right)$$

A similar model can be introduced for *y* and *z*:

$$y(x',y',z') = \sum_{n_{x'}=0}^{\infty}\sum_{n_{y'}=0}^{\infty}\sum_{n_{z'}=0}^{\infty} \frac{\partial^{n_{x'}} y}{\partial x'^{n_{x'}}} \frac{\partial^{n_{y'}} y}{\partial y'^{n_{y'}}} \frac{\partial^{n_{z'}} y}{\partial z'^{n_{z'}}} \frac{x'^{n_{x'}} y'^{n_{y'}} z'^{n_{z'}}}{n_{x'}!\,n_{y'}!\,n_{z'}!}$$

$$z(x',y',z') = \sum_{n_{x'}=0}^{\infty}\sum_{n_{y'}=0}^{\infty}\sum_{n_{z'}=0}^{\infty} \frac{\partial^{n_{x'}} z}{\partial x'^{n_{x'}}} \frac{\partial^{n_{y'}} z}{\partial y'^{n_{y'}}} \frac{\partial^{n_{z'}} z}{\partial z'^{n_{z'}}} \frac{x'^{n_{x'}} y'^{n_{y'}} z'^{n_{z'}}}{n_{x'}!\,n_{y'}!\,n_{z'}!}$$

Considering the measuring instrument to be in stationary conditions, partial derivatives can be considered to be couunstant:

$$\frac{\partial^{i+j+k} x}{\partial x'^i \partial y'^j \partial z'^k}\frac{1}{(i+j+k)!} = c_{xx'^iy'^jz'^k}\,; \quad \frac{\partial^{i+j+k} y}{\partial x'^i \partial y'^j \partial z'^k}\frac{1}{(i+j+k)!} = c_{yx'^iy'^jz'^k}\,; \quad \frac{\partial^{i+j+k} z}{\partial x'^i \partial y'^j \partial z'^k}\frac{1}{(i+j+k)!} = c_{zx'^iy'^jz'^k}$$

The following transfer function from instrument to metric coordinates can be consequently achieved:

$$\begin{bmatrix} x \\ y \\ z \end{bmatrix} = \begin{bmatrix} c_{xx'} & c_{xy'} & c_{xz'} \\ c_{yx'} & c_{yy'} & c_{yz'} \\ c_{zx'} & c_{zy'} & c_{zz'} \end{bmatrix} \cdot \begin{bmatrix} x' \\ y' \\ z' \end{bmatrix} + \begin{bmatrix} c_{xx'^2} & c_{xy'^2} & c_{xz'^2} & c_{xx'y'} & c_{xx'z'} & c_{xy'z'} \\ c_{yx'^2} & c_{yy'^2} & c_{yz'^2} & c_{yx'y'} & c_{yx'z'} & c_{yy'z'} \\ c_{zx'^2} & c_{zy'^2} & c_{zz'^2} & c_{zx'y'} & c_{zx'z'} & c_{zy'z'} \end{bmatrix} \cdot \begin{bmatrix} x'^2 \\ y'^2 \\ z'^2 \\ x'y' \\ x'z' \\ y'z' \end{bmatrix} +$$
$$+ \begin{bmatrix} c_{xx'^3} & c_{xy'^3} & c_{xz'^3} & c_{xx'^2y'} & c_{xx'^2z'} & c_{xx'y'^2} & c_{xx'z'^2} & c_{xy'^2z'} & c_{xy'z'^2} & c_{xx'y'z'} \\ c_{yx'^3} & c_{yy'^3} & c_{yz'^3} & c_{yx'^2y'} & c_{yx'^2z'} & c_{yx'y'^2} & c_{yx'z'^2} & c_{yy'^2z'} & c_{yy'z'^2} & c_{yx'y'z'} \\ c_{zx'^3} & c_{zy'^3} & c_{zz'^3} & c_{zx'^2y'} & c_{zx'^2z'} & c_{zx'y'^2} & c_{zx'z'^2} & c_{zy'^2z'} & c_{zy'z'^2} & c_{zx'y'z'} \end{bmatrix} \cdot \begin{bmatrix} x'^3 \\ y'^3 \\ z'^3 \\ x'^2y' \\ x'^2z' \\ x'y'^2 \\ x'z'^2 \\ y'^2z' \\ y'z'^2 \\ x'y'z' \end{bmatrix} + \sigma \begin{bmatrix} x'^3 \\ y'^3 \\ z'^3 \\ x'^2y' \\ x'^2z' \\ x'y'^2 \\ x'z'^2 \\ y'^2z' \\ y'z'^2 \\ x'y'z' \end{bmatrix}$$

The proposed transformation model can be implemented in order to describe 3D measuring instrument behaviour, with specific reference to systematic components. Conversely, stochastic contributes cannot be modelled through a mathematical function and are therefore comprised within σ Peano residual term.

IOP Conf. Series: Earth and Environmental Science **275** (2019) 012011 doi:10.1088/1755-1315/275/1/012011

Coefficients $c_{xx'}$, $c_{yy'}$ and $c_{zz'}$ are representative of linear calibration coefficients, while $c_{xy'}$, $c_{xz'}$ and $c_{yz'}$ quantify non orthogonality between instrument main coordinate directions. An ideally perfect scanner would exhibit values on the diagonal equal to 1, and null values out of the diagonal. Systematic non linearity along main axes are described by second and third order calibration coefficients: $c_{xx'^2}$, $c_{yy'^2}$, $c_{zz'^2}$, $c_{xx'^3}$, $c_{yy'^3}$ and $c_{zz'^3}$, while all of the other coefficients in the transformation matrix quantify second and third order coupled distortions. Estimation of different coefficients is not easy: however the aim of the model is not to provide a simple correction tool but, on the contrary, to introduce a general reference framework as a basis for instrument traceability.

2.3. Data filtering and extraction of parameters

The same TC 213 working on ISO 25178 created also an additional working group (namely WG 15) which actively contributed to the development of a toolbox containing different types of filters, eventually included within ISO 16610 [13]. The filtering operation is implemented for separation of surface components, such as roughness, waviness and form. Among the others, robust Gaussian, robust spline, morphological and wavelet filters can be applied. Robust filters are based on non-linear functions, which can generate mean surfaces not affected by local features or outliers such as high peaks or deep valleys. Morphological filters implement dilation and erosion operations, locally modifying features of the surface using a geometrical given shape called structuring element. Two operations are then defined: closing (a dilation followed by an erosion) or opening (dilation after erosion). Wavelet filters decompose surface into levels and allow extraction of components belonging to a specific scale level.

Filtered surface topographies can be eventually processed in order to allow computation and extraction of surface roughness parameters. ISO 25178 classifies 3D surface parameters into six main groups, as summarized in table 1.

Table 1. Main roughness parameters, as defined and classified by ISO 25178

Group	Abbreviated term	Notes	Applicative examples available at the		
			Micro-scale	Meso-scale	Macro-scale
height parameters	Sa, Sq, Ssk, Sku, Sp, Sv, Sz	Amplitude related methods, defined over the definition area	X	X	X
functional parameters (1)	Smr, Smc, Sk, Spk, Svk	Based on areal material ratio function of the scale-limited stratified functional surface	X	X	X
functional parameters (2)	Svq, Spq, Smq	Based on areal material probability curve, with the areal material area ratio expressed as a Gaussian probability	X		
volume functional parameters	Vm, Vv, Vmp, Vmc, Vvc, Vvv		X		
functional parameters	Svs, Srel, Svfc, Safc	Based on fractal cross-scale descriptive methods	X		
hybrid parameters	Sdq, Sdr	Defined on the basis of surface envelope and gradients	X	X	X
spatial parameters	Sal, Str	Based on autocorrelation functions.	X	X	X
miscellaneous parameter	Std	Based on texture direction of the scale-limited surface	X	X	X

The table shows also the present level of implementation of surface roughness parameters at a research or applicative level for the already cited different surface sizes: micro-, meso- and macro-scale (see e.g.

[14)]. The majority of applications are clearly found at the micro-scale, for millimetre and sub-millimetres surface analyses, but an increasing number of research studies is more recently implementing the same parameters at higher scales. Figure 3 reports the total number of published papers, as indexed by Scopus in the last decade and reporting roughness parameters in agreement with ISO 25178. The graph clearly shows how the number of documents is increasing at a rate higher than 30% per year. Such trend is mainly ascribable to the increasing recognition and understanding of the ISO 25178 parameters among both scientific community and private companies, and is reasonably expected to increase as an effect of the demand of simplified parameters for the analysis of surface data sets characterized by continuously growing resolutions.

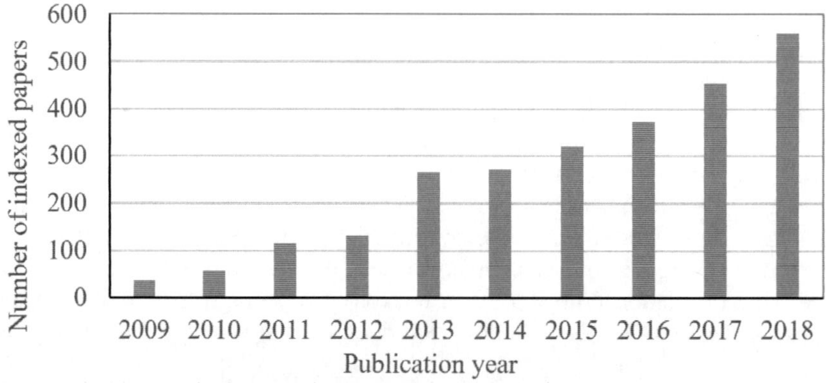

Figure 3. Number of published papers indexed by Scopus in the last ten years and implementing roughness parameters compliant with ISO 25178.

3. Conclusions

The present research is done with intention of producing an up-to-date understanding of the usage of surface roughness parameters at different scales.

Some conclusions can be drawn:

1. ISO 25178 provides an important reference for the definition of standard procedures and parameters for the characterization of surface topography data sets.
2. Specific filtering methods are also defined by the standard, allowing isolation of surface portions of interest, thus minimizing the uncertainty contributions related to noise, distortions or features not of interest for the specific applications
3. Specific calibration of data sets is needed in order to allow traceable computation of surface roughness parameters
4. In the point of view of the authors a growing number of research and applicative studies can benefit from extraction of parameters at different scales, from the micro- to the meso- to the macro-scale.

References

[1] McKean J and Roering J 2004. Objective landslide detection and surface morphology mapping using high-resolution airborne laser altimetry. *Geomorphology* **57** 331-351.

[2] Sofia G, Marinello F and Tarolli P 2015. Metrics for quantifying anthropogenic impacts on geomorphology: Road networks. *Earth Surface Processes and Landforms* **41** 240-255.

[3] Pezzuolo A, Guarino M, Sartori L, González L A and Marinello F 2018 On-barn pig weight estimation based on body measurements by a Kinect v1 depth camera. *Computer and Electronics in Agriculture* **148** 29-36.

[4] Pezzuolo A, Guarino M, Sartori L and Marinello F 2018 A feasibility study on the use of a structured light depth-camera for three-dimensional body measurements of dairy cows in free-stall barns *Sensors* **18(2)** 673

[5] Tosello G, Bissacco G, Tang PT, Hansen HN and Nielsen PC 2008. High aspect ratio micro tool manufacturing for polymer replication using μEDM of silicon, selective etching and electroforming. *Microsystem Technologies* **14** 1757-1764.
[6] Marinello F, La Storia A, Mauriello G and Passeri D 2018. Atomic Force microscopy techniques to investigate activated food packaging materials. *Trends in Food Science and Technology* in press.
[7] Dubbini M, Pezzuolo A, De Giglio M, Gattelli M, Curzio L, Covi D, Yezekyan T and Marinello F 2017. Last generation instrument for agriculture multispectral data collection *CIGR Journal* **19** 158-163
[8] Savio E, De Chiffre L and Schmitt R 2007. Metrology of freeform shaped parts. *CIRP Annals - Manufacturing Technology* **56** 810-835.
[9] Hocken RJ, Chakraborty N and Brown C 2005 Optical metrology of surfaces. *CIRP Annals - Manufacturing Technology* **54** 169-183.
[10] Grohmann CH, Smith MJ and Riccomini C 2011. Multiscale Analysis of Topographic Surface Roughness in the Midland Valley, Scotland. *IEEE Transactions on Geoscience and Remote Sensing* **49** 1200-1213.
[11] ISO 25178. Geometrical Product Specifications (GPS) – Surface Texture: Area – Part 2: Terms, Definitions and Surface Texture Parameters, International Organisation for Standardization, Geneva, 2012.
[12] Marinello F; Bariani P; Carmignato S and Savio E 2009. Geometrical modelling of scanning probe microscopes and characterization of errors *Measurement Science and Technology* **20** 084013.
[13] ISO 16610. Geometrical product specifications (GPS) - Filtration - Part 21: Linear profile filters: Gaussian filters, International Organisation for Standardization, Geneva, 2011.
[14] Todhunter L.D., Leach R.K., Lawes S.D.A. and Blateyron F. 2017. Industrial survey of ISO surface texture parameters *CIRP Journal of Manufacturing Science and Technology* **19** 84-92.

IOP Conf. Series: Earth and Environmental Science **275** (2019) 012012 doi:10.1088/1755-1315/275/1/012012

Analysis of the effects of uncertainties on agrometeorological models

A. Garinei[1], A. Marini[2], E. Piccioni[3], M. Marconi[1], L. Biondi[1], A. Bertolini[4] and G. Ferrari[4]

[1] Guglielmo Marconi University, Department of Sustainability Engineering, via Plinio 44, 00193, Roma, Italy
[2] INFN Section of Perugia, Via A.Pascoli, 06123, Perugia, Italy
[3] Idea-re S.r.l., Via Cornelia, 498, 00166, Roma, Italy
[4] Radarmeteo S.r.l., via IV Novembre 119, Due Carrare, 35020, Padova, Italy

E-mail: a.garinei@unimarconi.it

Abstract. We analyze how the uncertainties in meteorological data affect the outcome of an agrometeorological model. In particular we focus on a model simulating the development of primary infections of downy mildew on grapevine. This model takes as inputs hourly measurements of temperature, relative humidity and rainfall and it simulates all the phases that eventually lead to the downy mildew primary infections. In order to assess the robustness of the model against the uncertainties of the input data we set up a Monte Carlo simulation. This procedure enhances the potentiality of the model since allows to extract, as an additional valuable piece of information, the probability of the infection event. The results we obtain are highly significant and suggest that the effects of the uncertainties are actually important and should be carefully evaluated.

Keywords: Uncertainties in Meteorology, Agrometeorological Models, DSS.

1. Introduction

Smart techniques applied to the agricultural sector are becoming increasingly spread and important. In this scenario, agrometeorological models are formidable tools in farmers' hands which can provide a helpful guidance in taking timely decisions and consequently boost the crop productivity. Naively, the outcomes of these models are heavily dependent on the input data, both in term of the amount of data available and of their quality. In many cases, it is reasonable to expect that even relatively small changes in the input data could drastically change the outcome of the model. So, considering that each measurement is affected by a certain amount of uncertainty, a valuable question that we can ask is whether and to what extent we can trust the results of a model. Despite being crucial, this aspect is usually neglected in real implementation of these models and in their actual use in the agricultural sector. Currently many of the agrometeorological models are based on some internal deterministic dynamics and, once fixed the input data, they yield a certain and unique output; there are no shades in the results. If one took into account the uncertainties of the inputs and determined how they propagate, it would be possible to enhance the amount of information that can be extracted from these models: namely the latter could tell not just if a certain event will happen but rather what is the odd that it will happen. This could make such models more useful and also make easier the integration of the results with other sources of information. For example, in the context of decision support systems (DSS), the knowledge of the probability that a certain disease may be present in a field could be integrated by the farmer experience and help in taking the decision to start the suitable phytosanitary treatment. Note furthermore, that this kind of investigations could be extremely interesting and have relevant applications in the agricultural insurance field: for instance it could provide the guidelines on the good practices giving access to the benefits offered by mutual aid societies when natural events or disasters harm the crop productivity.

In this paper, we analyze from a quantitative point of view how the uncertainties in the input data affect the outcome of an agrometeorological model. In particular, we focus on a model simulating the development of primary infections of downy mildew on grapevine, proposed by Rossi et al. [1, 2]. Downy mildew is considered the most devastating disease affecting grapevines and it is caused by a heterothallic oomycete, the plasmopara viticola [3]. During the wintertime, this oomycete lives in the soil in the form of oospores, which in spring germinate and spawn macrosporangia. Under wet conditions, macrosporangia release zoospores, which can be dispersed and eventually reach grapevine's leaves through rain splashes causing primary infections on plants. The infection becomes manifest after few days when yellow lesions appear on leaves. During favorable weather, these lesions can sporulate and also cause new secondary infections.

The agrometeorological model we consider in our study takes as inputs hourly measurements of temperature, relative humidity and rainfall and it simulates all the phases that eventually lead to the downy mildew primary infections. We start from a set of real meteorological data, collected by a weather station and we set up a Monte Carlo simulation in order to assess the robustness of the model against the uncertainties of the input data. The simulation is carried out by adding statistical noise to the real data with amplitudes determined according to the level of uncertainty affecting the measured data. This is given by the measurement error, which is determined by the combination of the instrumental error and the siting error, induced by the environment surrounding the weather stations. Both these two sources of error can be estimated on the basis of the compliance classification made by the World Meteorological Organization (WMO) [4, 5, 6]. Running the simulation several times we were able to determine the impact of the uncertainties on the model outcome for all the possible compliance classes.

We also perform an analogous analysis in the case in which the input data that feed the agrometeorological model are not measured directly from a physical weather station located in the field, but rather they are determined by means of geostatistical procedures. Indeed, weather forecast providers have highly sophisticated services that allows to retrieve local meteorological data even when there are no stations in the fields, using data collected by surrounding stations and interpolating these data through geostatistical methods. In such a way, one can reconstruct punctual meteorological data for every location in the territory covered by the network of weather stations. In this case the uncertainties of the data are generally higher since, besides the measurement errors, one has to consistently take into account also the statistical error introduced by the interpolation procedure.

2. Materials and methods

We consider as a case study an agrometeorological model for the development of primary infection of plasmopara viticola (downy mildew) on grapevine [1, 2]. The model uses hourly measurements of temperature, relative humidity and rainfall as inputs to simulate in a mechanistic way the dynamics that leads to downy mildew primary infections. The model's dynamics is quite involved but for our scope a full understanding of how it works is not required; we refer the interested reader to the original articles for a detailed description.

The aim of our study is to assess the robustness of the model against the uncertainties of input data. We consider two different scenarios: the first in which the weather data come from

Table 1. Maximum instrumental errors for each of the three classes according to the WMO classification. The errors for temperatures and that for rainfalls below 5mm are absolute while the ones for relative humidity and rainfalls above 5mm are relative (expressed as percentage).

	Temperature (°C)	Relative Humidity (%)	Rainfall if < 5 mm (mm)	if > 5 mm (%)
Fully compliant (class 1)	0.2	3	0.3	5
Compliant (class 2)	1	10	0.5	10
Non-compliant (class 3)	2	20	1	20

direct measurements taken by stations installed in the field we want to monitor; the second one where the weather data are obtained through geostatistical interpolations of data coming from surrounding stations.

2.1. Data from real stations

We considered six datasets of meteorological measurements collected form weather stations in Veneto, a region of Italy particularly devoted to viticulture and renowned in the world for its wines, like Prosecco and Amarone. More precisely, we used data from the following six locations:

- Malo (Vicenza);
- Tribano (Padova);
- Illasi (Verona);
- Marano di Valpolicella (Verona);
- Brendola (Vicenza);
- Conegliano Veneto (Treviso).

The time window covered by the datasets spans from the January 1st to August 31th, 2017, in accordance to what required by the model [1, 2]. Figure 1 shows the behavior of the measured data as a function of time (day of the year).

Starting from the measured data we set up a Monte Carlo procedure aimed to the derivation of a statistical evaluation of the uncertainties affecting the outcome of the model. For each measurement, we generated a random variable having the actual measured value as mean and the measurement error as standard deviation. In particular we used Gaussian distributions for all the three input variables, namely the temperature, the relative humidity and the rainfall. Since, naively, relative humidity is allowed to be between 0 and 100 and rainfall to be strictly non-negative we also imposed these constraints on the generated variables. We considered two contributions to the measurement error associated to the three meteorological quantities used as inputs in the model: the instrumental and the siting errors. The instrumental error is the one associated to the accuracy of the measurement devices installed in the station. According to the WMO classification we considered three classes: fully compliant, compliant and non-compliant instruments. The maximum instrumental error associated to each class follows the WMO specifications [4, 5] and it is reported in Table 1.

The measurement error is not only determined by the quality of the instruments installed but also by the location of the station. The environmental conditions can indeed influence the outcome of the measurements. The WMO defined a set of relevant criteria to assess the goodness of the sites and, according to these criteria, it provided a fivefold classification, where class 1 corresponds to an optimal location and class 5 to the worst environmental condition for

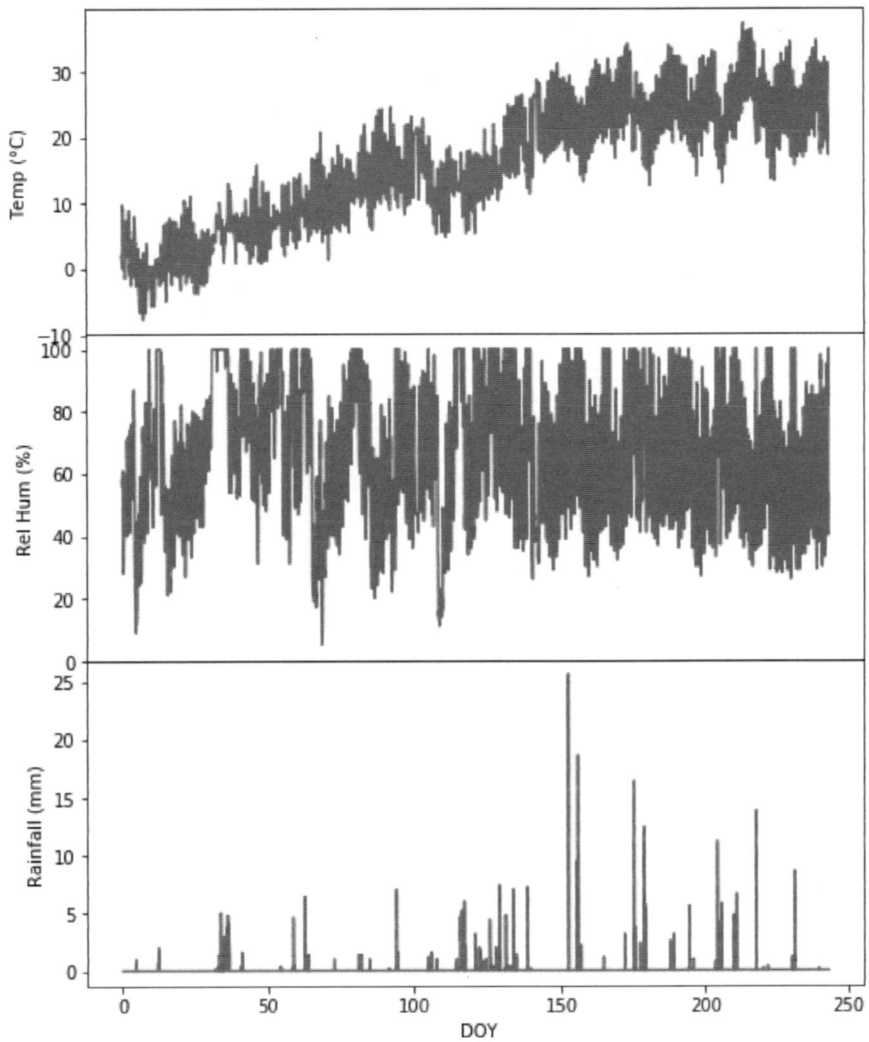

Figure 1. Temporal behavior of the physical quantities used as inputs in the model as measured from the station located in Malo. Horizontal axis: the day of the year (DOY); vertical axis: temperature, relative humidity and rainfall.

the weather station [6]. Table 2 shows the siting induced errors for temperature and rainfall measurements for each of the five classes. The errors for relative humidity measurements are not given so we will assume that they are not significantly influenced by the environmental conditions.

In the Monte Carlo simulations, we will consider different combinations of instrumental and siting classes, fixing the uncertainties to the maximum value of the error for the considered class, according to Tables 1 and 2. The instrumental and siting errors are then combined in quadrature yielding the overall uncertainty associated to the measurement.

Table 2. Maximum error for each site class according to the WMO classification. The error for the temperature is absolute while the one for rainfall is relative (%).

	Temperature (°C)	Rainfall (%)
Class 1	–	–
Class 2	–	5
Class 3	1	15
Class 4	2	25
Class 5	5	100

2.2. Data from virtual stations

Besides the analysis explained in the previous section, we consider also the case in which the data used as inputs of the agrometeorological model are not measured directly by devices located in the field, but rather are determined by means of geostatistical interpolations [7, 8]. This is a particularly relevant case, since it is quite common that there are no weather stations installed on or nearby the field one wants to monitor. In such situations, one can still retrieve the needed data using the services provided by companies operating in the weather forecast sector. Indeed, usually these companies have access to the meteorological data from a wide network of stations spread on the territory and thus they can use geostatistical interpolations to derive an estimate of the variables at any locations. Since in this case there is not a real physical station installed in the field, we say that the data comes from a "virtual station".

We chose to apply the model in the same time period and in the same six positions considered in the previuos section, using this time the "virtual" measurements instead of the real ones recorded by the physical stations. Of course, in each case we pretended that there was not a weather station in that place, *i.e.* in the interpolation procedure carried on to derive the "virtual" measurements we discarded the data from the real station located in that place. Specifically, the data are obtained by means of an interpolation technique particularly suited for this kind of applications, called kriging method [9, 10, 11].

Figure 2, analogously to what displayed in Figure 1 for the real station, shows the temporal behavior of the input variable of the models as "measured" by the virtual station.

The procedure we use in order to estimate the uncertainties induced on the model outcome by the data is the same as for the previous case of data collected from real station. What changes in this case is the error associated to the input data, since we now have to take into account also the contribution coming from the geostatistical interpolation itself. The statistical error associated to the interpolated data has been estimated applying the leave-one-out cross-validation procedure to the measurements coming from the set of N stations involved in the interpolation [12, 13]. In practice, for each of these stations one estimates the meteorological data applying the kriging method, using the data from the remaining $N - 1$ stations. Then for every physical quantity, one defines an associated "error field" whose value in each station location is given by the absolute value of the difference of the real measurement and the estimated one. Finally, using again the geostatistical interpolation applied to these error fields one can estimate the values of uncertainties at the location of the virtual station. Notice that this procedure yields uncertainties which are specific for each location and, differently from the ones associated to the real measurements, variable in time, since the cross validation is carried out independently for each hourly measurement. The statistical error due to the geostatistical procedure is then combined with the measurement errors associated to the real measured data

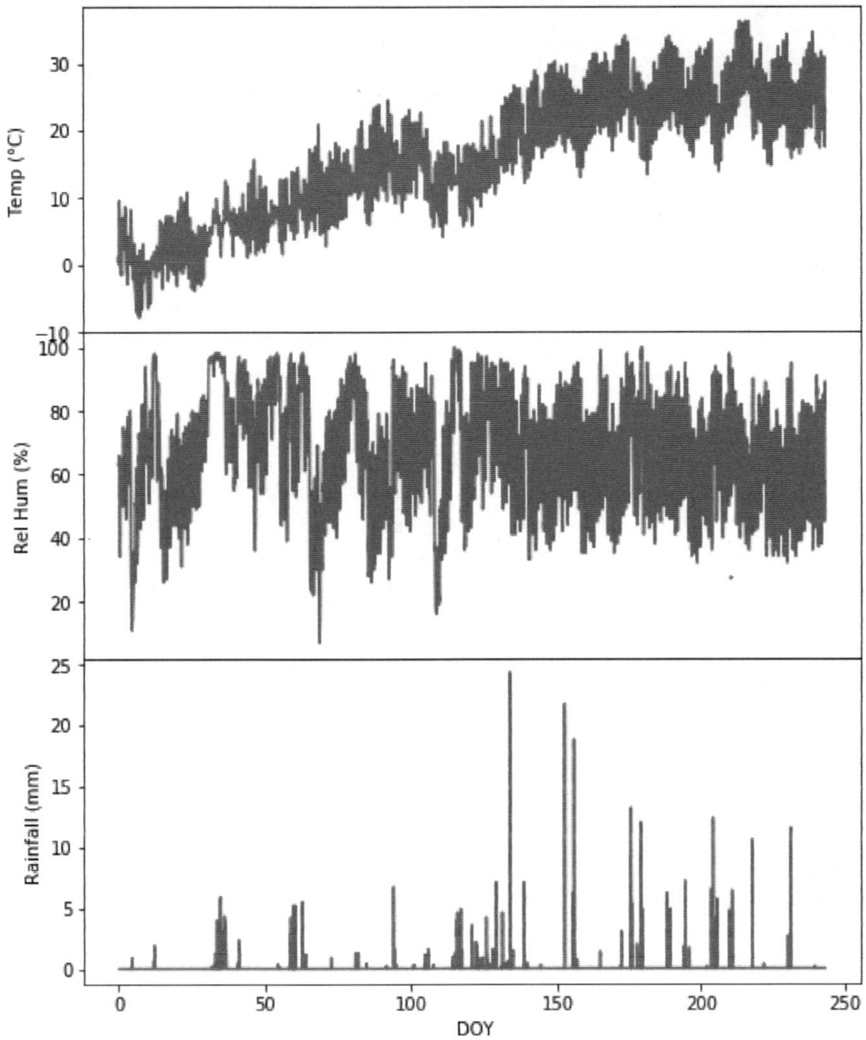

Figure 2. Temporal behavior of the physical quantities used as inputs in the model as obtained from the virtual station in Malo: on the horizontal axis, the day of the year (DOY) is reported, and on the vertical axis temperature, relative humidity and rainfall.

used in the interpolation. In principle, each measured data has its own error which is related to the compliance of the station from which it is taken, namely to the instrumental specifications and to location of the station, as discussed in Section 2.1. However, to keep the things simple, we assume that all the stations used to estimate the virtual data are of the same compliance class and we explore how the results of our analysis change by varying the compliance classes.

3. Results

As a first step, we just run the agrometeorological model using the "real" data, namely the ones measured by the instruments physically installed in the locations considered in this study. An example of the output of the model, for the station located in Malo, is summarized in Figure 3, showing the temporal behavior of the various phases of disease development.

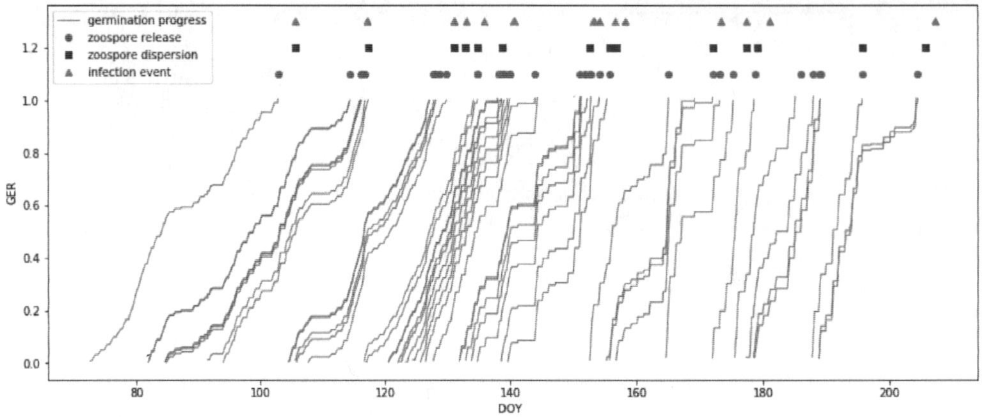

Figure 3. Plot summarizing the progress of the various phases that lead to the development of the disease as a result of the model run over measurement from the real station in Malo. The day of the year is represented on the horizontal axes; the lines indicate the germination progress (GER – ranging from 0 to 1) for each cohort of plasmopara viticola; green circles correspond to zoospore release events, blue squares to zoospore dispersion events, red triangle to infection events.

We focus only on the infection events (depicted as red triangle in the plots in Figure 3) whose prediction is indeed the main goal of the model and clearly it is what really matters for agricultural purposes. The main goal of our research is to derive an indicator of the reliability of the outcome of the model in relation to the uncertainties affecting the data used as inputs. We choose to use the probability that a downy mildew primary infection starts at a given day as such indicator. This is obtained using Monte Carlo simulations: we run the model several times using generated data and the probability of infection at a given day is computed as the fraction of simulations which predicted the onset of an infection on that day. In particular we fixed the number of runs, N, to one thousand, since we observed that this number guaranteed a good stability of the outcomes, *i.e.* further increasing N did not have significant effect on the results of the simulation.

Note that the model we use takes as inputs hourly data and gives outputs with hourly resolution as well. However, in the computation of the probabilities, we decided to recast the output on a daily grid in order to avoid an excessive complexity of the model and to make our results more stable, reducing the chance that a simple few-hours translation of an event in different runs is interpreted as a discrepancy. This is also motivated by the fact that in practical uses in agriculture it is sufficient to know the response on a daily level.

3.1. Data from real stations

As explained before, when the meteorological data necessary in order to run the model come from measurements made by weather station installed in the field, the uncertainties that affect the outcome are due to the instrumental and siting errors. We considered different combinations of instrumental compliance and siting classes, according to the classification reported in the previous section. For ease of notation, we denote a station belonging to the instrumental compliance class n_i and to the siting class n_s simply through the ordered pair (n_i, n_s). For the sake of brevity we report here only some of the results we obtained: focusing again on the station in Malo, the plots in Figure 4 show the results for the estimated infection probabilities in the cases of compliance classes (1,1), (1,3), (2,2) and (3,3).

We notice that when the instruments in the station are fully compliant ($n_i = 1$) and optimally located ($n_s = 1$), *i.e.* when the uncertainties associate to the measurements are the least possible, the probabilities associated to the predicted infection events are very clearly polarized in almost all the cases. This means that results provided by the model are quite robust. As one naively expects, the peaks in the probabilities tend to lower and spread out as we take into account lower quality compliance classes, namely as we increase the uncertainties.

It is interesting to note that even if consider a case of a mid-level quality station, for instance in the class ($n_i = 2$, $n_s = 2$), the information that our procedure adds to the simple and dichotomous output of the model becomes in some cases really valuable, as clearly shown in the third plot of Figure 4.

3.2. Data from virtual stations

We now repeat the same analysis as before using data coming from interpolations rather than from direct measurements in the field. Firstly, in Figure 5 we report the output of the model run over the interpolated data for the virtual station located in Malo.

Comparing the results obtained using the real (Figure 3) and virtual (Figure 5) data for the station in Malo we notice that, the two outputs are quite similar, though not exactly coincident. Repeating the same comparison for the other locations considered in our study we found that in some cases (in particular Illasi and Marano) the discrepancies are sharper. This is a hint that the agrometeorological model we are considering is quite sensitive to the precision of the input data and in some cases the interpolated measurements may be not so reliable for the required task.

Then, as we did in our analysis for real stations, we run the Monte Carlo procedure to determine the robustness of the model, considering different combinations of instrumental and siting classes (n_i, n_s) for the network of stations involved in the interpolation. Figure 6 shows some examples of the results of the Monte Carlo simulation, for compliance classes (1,1), (1,3), (2,2) and (3,3). Note that, differently from what happens for real stations, in this case the results for different compliance classes are almost identical, since for virtual station the dominant contribution to the uncertainties comes from the interpolation error.

4. Conclusions

We developed a procedure to assess the robustness of an agrometeorological model against the uncertainties that affects the input data. These uncertainties cannot be completely removed since they are related to the physical characteristics of the measurement devices installed in the stations (instrumental sensitivities) and to environmental factors of the surrounding area. We considered a specific model as a case study, namely a mechanistic model for the development of primary infections of downy mildew on grapevine. The results that we obtained are highly significant and suggest that the effects of the uncertainties are actually important and should not be neglected.

The procedure we used to estimate the propagation of the uncertainties consisted essentially in a Monte Carlo simulation repeated several times with generated data. At every run, the model gives as outputs the predicted downy mildew infection events. These outputs allowed us to derive an estimation of the probability of the onset of an infection at a given day of the year, simply by counting the fraction of simulations yielding that result. It should be noted that in order to generate the data for the simulations we had to make some strong assumptions on their statistical distributions. We considered in particular normal distributions for all the measured input variables, with mean on the actual measured value (or the interpolated value resulting from the kriging procedure in the case of virtual stations) and variance determined according the estimated error of the measurement. This assumption, although being standard, is

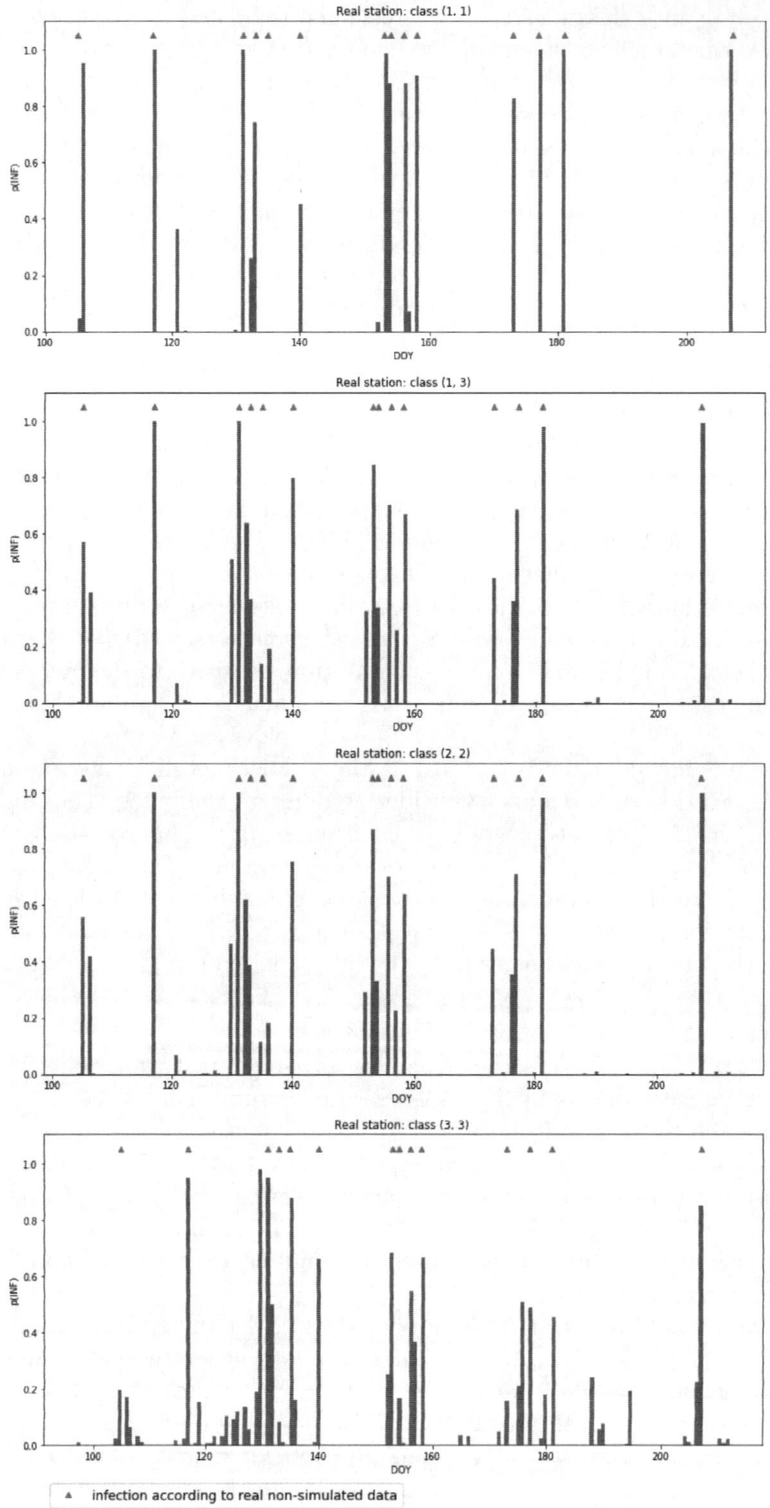

Figure 4. Plots showing the probability of the onset of infection at a given day of the year (DOY) for the real station located in Malo in the cases of compliance classes (1,1), (1,3), (2,2) and (3,3). The red triangles in the upper part indicate the days of infection according to the model run with the original (non-simulated) real data.

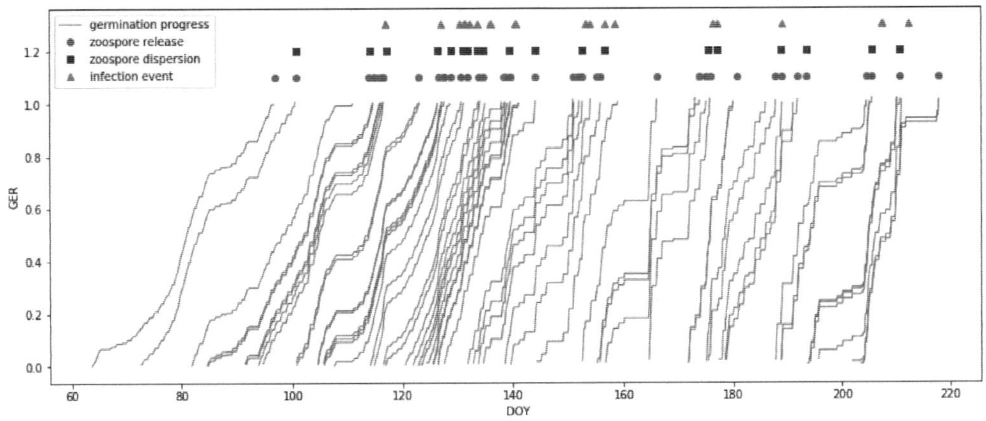

Figure 5. Plot summarizing the progress of the various phases that lead to the development of the disease as a result of the model run over measurement from the virtual station in Malo. The day of the year is represented on the horizontal axes; the lines indicate the germination progress (GER – ranging from 0 to 1) for each cohort of plasmopara viticola; green circles correspond to zoospore release events, blue squares to zoospore dispersion events, red triangle to infection events.

quite strong: it implies, for instance, that we are considering measurements that are on average unbiased and no systematic error is present. This is a caveat that one has to take into account when interpreting the results.

Even if in this paper we considered only a specific agrometeorological model, the procedure we implemented to estimate the propagation of the uncertainties can be applied to any model. Besides the actual results we showed in Section 3, the main and most relevant objective of the paper has been to introduce a new general paradigm of implementation of agrometeorological models and demonstrate that it really increases the information that can be extracted from the models and allows the user to have a clearer interpretability of the outcomes.

We decided to apply the aforementioned procedure in an extremely involved agrometeorological model, which takes as inputs a big amount of data (hourly measurements of temperature, relative humidity and rainfall for an extended time window) and elaborate them in a highly non-linear fashion in order to produce the desired output. Indeed, that of complicated models is exactly the case in which one can expect some chaotic behaviors to emerge when perturbing the input data and it is thus the most interesting instance to consider.

Finally, it is worth noting that the obtained results suggest that one needs to be very cautious when using data from virtual stations since these may be not so reliable in the application to a particular agrometeorological model. Therefore, when one uses such "virtual" data, a careful evaluation of the actual effectiveness of the procedure for the specific case at hand is highly recommended.

Acknowledgments

We thank *Radarmeteo* company for providing the data used for our elaborations. The study is part of a work carried out in the framework of the *RESA project*, funded by *Regione del Veneto – POR FESR 2014-2020*.

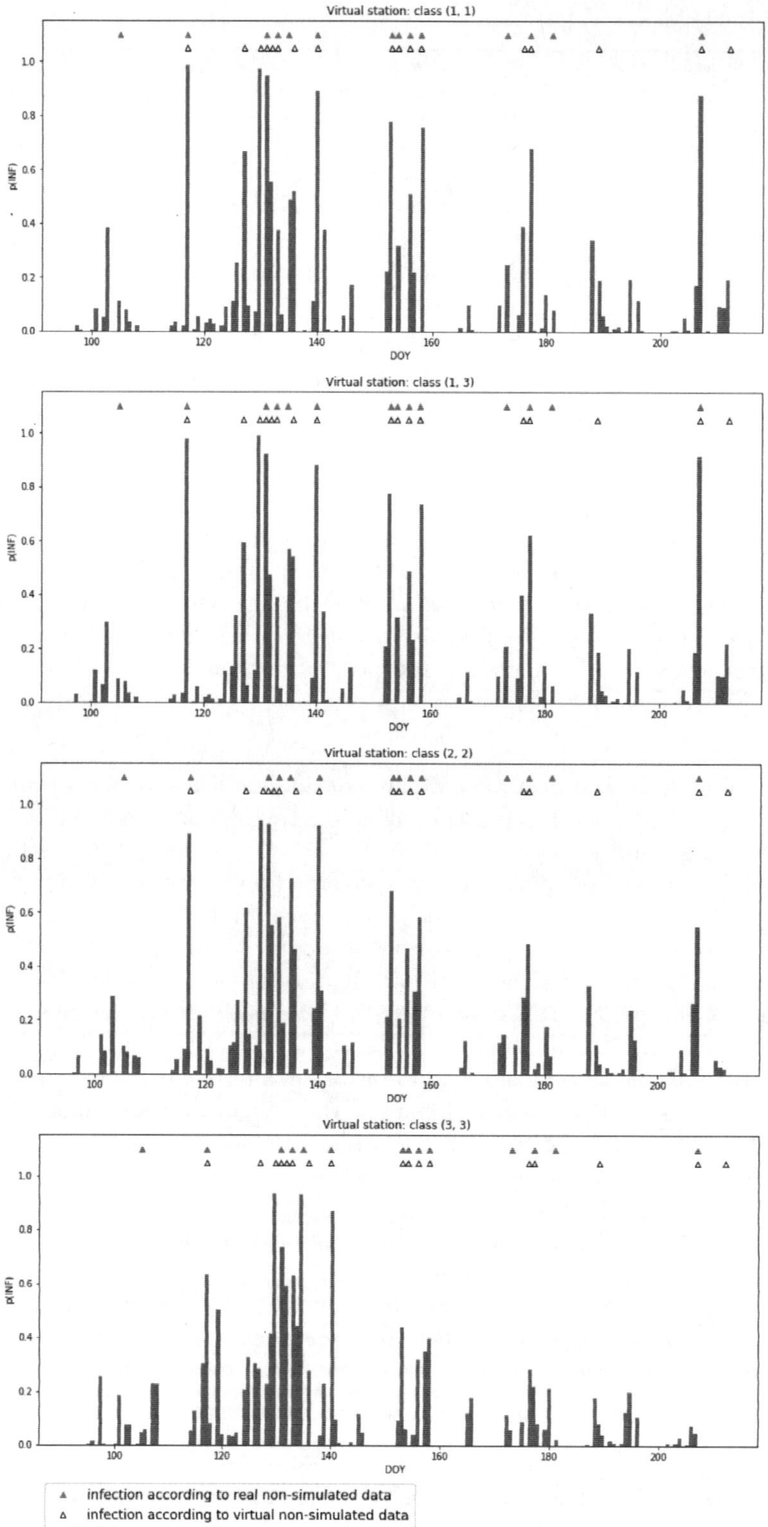

Figure 6. Plots showing the probability of the onset of infection at a given day of the year (DOY) for the virtual station located in Malo in the cases of compliance classes (1,1), (1,3), (2,2) and (3,3). The red and black (empty) triangles in the upper part indicate the days of infection according to the model run with the original (non-simulated) real and virtual data, respectively.

IOP Conf. Series: Earth and Environmental Science **275** (2019) 012012 doi:10.1088/1755-1315/275/1/012012

References

[1] Rossi V, Caffi T, Giosuè S and Bugiani R 2008 *Ecological Modelling* **212** 480–491
[2] Caffi T, Rossi V, Bugiani R, Spanna F, Flamini L, Cossu A and Nigro C 2009 *Journal of Plant Pathology* 535–548
[3] Gessler C, Pertot I and Perazzolli M 2011 *Phytopathologia Mediterranea* **50** 3–44
[4] WMO 2008 *Guide to meteorological instruments and methods of observation* WMO-No. 8 (Geneva: World Meteorological Organisation)
[5] Fisler J, Kube M, Stocker C and Grüter E 2017 *Quality assessment using METEO-Cert: The MeteoSwiss classification procedure for automatic weather stations* (World Meteorological Organisation)
[6] WMO 2010 *Commission for instruments and methods of observation – Report of the fifteenth session* WMO-No. 1064 (Helsinki: World Meteorological Organisation)
[7] Bivand R S, Pebesma E J, Gomez-Rubio V and Pebesma E J 2008 *Applied spatial data analysis with R* (*Use R* vol 10) (New York: Springer)
[8] Diggle P and Ribeiro P J 2007 *Model-based Geostatistics* Springer Series in Statistics (New York: Springer)
[9] Matheron G 1963 *Economic geology* **58** 1246–1266
[10] Cressie N 1990 *Mathematical geology* **22** 239–252
[11] Sluiter R 2009 *Interpolation methods for climate data* (KNMI)
[12] Pebesma E J 2004 *Computers & Geosciences* **30** 683–691
[13] Pebesma E J and Wesseling C G 1998 *Computers & Geosciences* **24** 17–31

IOP Conf. Series: Earth and Environmental Science **275** (2019) 012013 doi:10.1088/1755-1315/275/1/012013

Data analysis and inference model for automating operational monitoring activities in Precision Farming and Precision Forestry applications

Pasqualina Sacco[1], Raimondo Gallo[2], Fabrizio Mazzetto[2]

[1]Fraunhofer Italia Research scarl, Via Volta 13A, 39100 Bolzano (Italy)

[2]Faculty of Science and Technology, Free University of Bozen (FUB), Piazza Università 5, 39100 Bolzano (Italy)

Corresponding author: Pasqualina Sacco, pasqualina.sacco@fraunhofer.it

Abstract. Each application of Precision Agriculture or Forestry should be supported by a technological platform able to perform, in an integrated way, the following data-information cycle functions: 1) data collection; 2) data processing; 3) data analysis and evaluation; 4) use of information. In accordance to this view, information are data that are usefully used in a decision making process or within a reporting protocol destined to users external to the enterprise (certification tasks). In order to manage the platform in a complete and efficient manner an adequate information system is needed. Firstly, the paper shows a classification of the possible monitoring solutions based on the different enterprise typologies, highlighting the main technological and interpretative requirements. Secondly, some case studies related to the application of operational monitoring in orchards and forestry are introduced, mainly focusing on some peculiar aspects of the algorithms developed for the implementation of the inference engines.

1. Introduction

The evolution of ICT in production processes in industry sector is well known and leads to the industry 4.0 concept. By analogy we hear about agriculture 4.0, smart agriculture or similar definitions, but despite them the fundamental concepts that connect Industry 4.0 and Precision Agriculture are: 1) automation; 2) hyper-connectivity among production components and devices (cybernetic approaches); 3) treatment of huge amount of data (Big Data); 4) quick interpretation of data achieved for targeted and quality decision making processes.

Each application of Precision Agriculture or Forestry should be supported by a technological platform able to perform, in an integrated way, the following data-information cycle functions: 1) data collection; 2) data processing; 3) data analysis and evaluation; 4) use of information. In accordance to this view, information are data that are usefully used in a decision making process or within a reporting protocol destined to users external to the enterprise (certification tasks). In order to manage the platform in a complete and efficient manner an adequate information system is needed, capable of satisfying the specific organizational needs of agro-environmental enterprises [1, 2].

Such an information system has to be designed following an info-logical approach, with the following characteristics: 1) considering the entire data-decision cycle from the design of the individual phases (e.g. data acquisition targeted to the decision-making objectives); 2) using hardware and software technology to solve technical and cultural aspects (e.g. attention to the presentation of information); 3) break down the objectives into simple and modular task to provide solutions that are adaptable to different contexts (e.g. environment, production goal, dimension); 4) break down the objectives according to the different decision levels (strategical, tactical, operational).

Data collection function requires relevant technological and organizational efforts, for the multiplicity of aspects to be kept under constant monitoring, and its whole management system must be designed based upon a decision-related view, rather than a data-related view. In particular, operational monitoring is the most complex from an organizational standpoint, especially when it is expected to be carried out in an automated way, because it involves all the functions of the whole data-information cycle. By definition, operational monitoring includes tasks performed to survey, record and report the information necessary to get an overview of the execution details applied in a given process activity [3]. To this standpoint, it must be distinguished from other monitoring functions such as environmental and crop monitoring.

Figure 1 shows a possible classification kay of monitoring solutions based on the different enterprise typologies, highlighting the main technological and interpretative requirements. In fact, the object of operational monitoring varies according to the farming system (arable farms, orchards, livestock, forestry enterprises, etc.) and the main decision-making processes that must be there undertaken. In general, the automation of operational monitoring requires: a) different data acquisition systems, with characteristics conditioned by the dynamic modalities with which the operation at hand is carried out; b) an inference engine, as an integral part of the processing procedures, to synthetically interpret the relevant aspects of the work carried out; c) an access system to information, to allow an interactive evaluation, as well as a correct use of information depending on whether the decision-making processes concern aspects internal to the enterprise (control and organization of activities) or external to it (certification, traceability). With reference to Figure 1, for example, the geographical monitoring referred to the "site of work" characteristic is less important for livestock farms (intensive livestock farming) than the others. However, the use of information must be quick to adapt to the stable management processes.

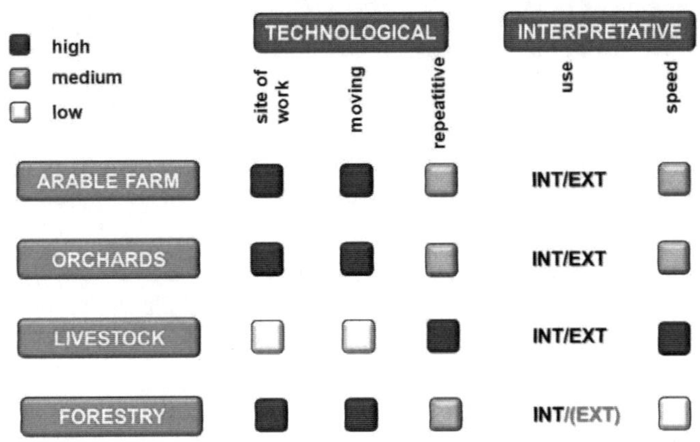

Figure 1. Monitoring solutions based on the different enterprise typologies and the main use of the information in decision making processes.

2. Farm and forestry information systems application

In this section, some case studies related to the application of operational monitoring in orchards and forestry are introduced, focusing on some peculiar aspects of the data acquisition devices there applied and the algorithms developed for the implementation of the inference engines.

2.1. Operational monitoring in orchards

In the case of orchards, the objective is to automatically reconstruct the sequence of work events in the field by identifying: a) type of operation performed; b) machines involved; c) workplaces (crop and/or varieties treated). The project focused on mechanized field activities and the automatic fill of the field activities performed [3]. This is done providing both a summary of the way the operation at hand was carried out - with details on each work phase (actual and auxiliary times, including stops, supplies and intra- and extra-farm transfers) - and an estimate of fuel consumption.

The transition from data to information is transparent to the end user. Everything is charged to the analysis algorithms that deal with processing the new data as they arrive and prepare the summary information ready to be evaluated and used (Figure 2).

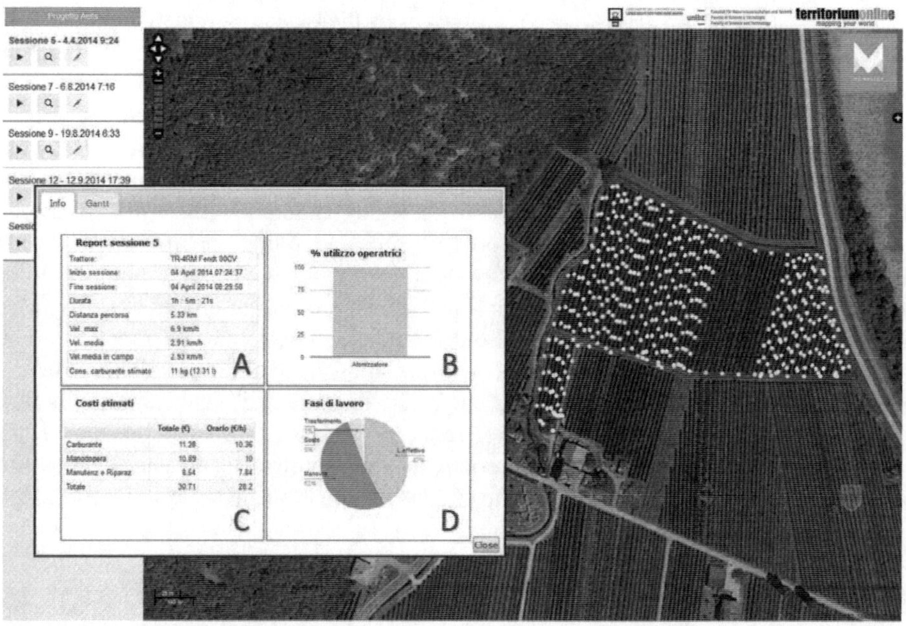

Figure 2. Example of user interface showing details of a single farm activity, as output of an operational monitoring system. A: detail of the selected working session; B: use of implements in percentage of total time; C: estimated costs; D: details of elementary working phases (effective work, stop, manoeuvres, and transfers).

Concerning data acquisition, there are several possible constructive architectures of data acquisition systems for the automation of operational monitoring (Figure 3). The main difference is between the position of the data logger that could be tractor-on board or implement-on board. The solutions A, B and C are of the tractor-oriented type, with the acquisition devices installed on the tractor. The solution D is implement-oriented, having the devices installed directly on the implement. Solution C is equipped with an identification system installed on implements for the autonomous recognition of the operations performed [4]. The implement-on board solution gives more details of the operation but is able to follow one operation at a time. The tractor-on board solution is more flexible, but requires the capability of identifying the implement.

IOP Conf. Series: Earth and Environmental Science **275** (2019) 012013 doi:10.1088/1755-1315/275/1/012013

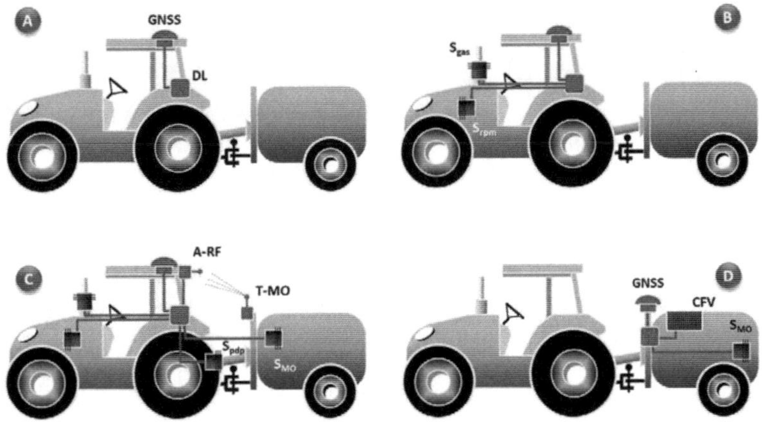

Figure 3. Data acquisition systems for the automation of operational monitoring [4]. DL = data logger; GNSS = receiver for satellite positioning system; SX = sensors for aspects of tractor operation or the operator; T-MO = transmitter of identification code of the operating machine (in RF); A-RF = receiver that identifies the transmitted codes; CFV = photovoltaic cells.

Our tests in orchard were performed implementing architectures mainly referred to solutions of type B (forest monitoring) and C (monitoring in vineyards and orchards). More in detail (Figure 4), field data logger identify the power unit (tractor), and are equipped with: accelerometer, positioning system (GPS+GLONASS receiver), transmission system (GPRS-UMTS modem unit, real time), RF receiver (to log data from coupled implement) and internal memory buffer. Implements are equipped with active tag to identify the implement it self, accelerometer to automatic turn on of the tag (active only when the implement is coupled and moving), RF transmitter (low power signal emitter, 433 MHz - UHF, broadcasting distance < 10 m), long life battery (approx. 500 hrs).

Figure 4. Data acquisition systems details. From the left the first two images show the data logger on the tractor, the third image shows one of the code transmitter on an implement [3].

Once data are acquired, the inference engine extracts the maximum information from all available data and combines it in order to obtain the final information on the activities carried out with the highest degree of reliability possible. Monitoring tractor's activity two different behaviour could be considered: kinematic and functional. Kinematic behaviour identifies the tractor's activity from position data; it comes to identifying the working pattern, but it is difficult to infer the operation performed. Functional behaviour identifies the tractor activity from additional information related to specific properties of the operation in progress (e.g., functional site map, implement code transmitters). The degree of reliability of the inference depends on the data availability. The minimum equipment is the farm monitoring system (data logger on tractor). Kinematic is always good inferred, but activity inference reliability depends on other information availability to refine the kinematic model.

The whole computational process (Figure 5) includes a mixed of fuzzy logic, statistic and spatial analysis algorithms, and generally works on the assumption that the farm map is a priori known. This map includes both elementary shape areas (say a cultivation unit, CU, set as polygon assigned to a specific crop or variety; one or more CUs form a field) and specific functional points of interest (downloading/uploading sites, refilling points, workshops, shelters etc.). Anyhow, if a map is not a priori available (e.g. in case of contractors), additional sets of spatial algorithms permit to identify areas where works were carried out following special patterns (e.g., works with parallel adjacent rows or alternating rows, transport cycles to the farm centre).

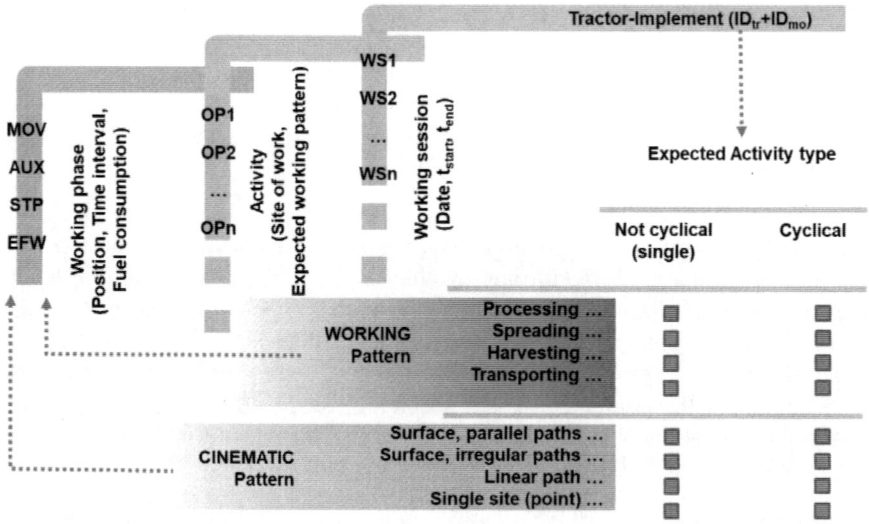

Figure 5. Inference model for detecting mechanized field activity on farm [3].

Figure 6 shows an example of application of the inference engine for the operational monitoring of anticryptogamic treatments in apple orchards. The objective is to interpret the types of activities carried out in a work session (SL) together with the related execution methods (work phases, times and fuel consumption). Every single fixing represents the minimum elementary state (SE) of the activity and the sequence of the SEs on the map shows the entire route taken. Each SE is first classified as actual work, stop or transit based on parameters that evaluate the instantaneous values of speed and direction. A further analysis based on spatial contiguity with other SEs fixes the portions of space in which parallel contiguous passages have been made. A subsequent clustering then identifies the related areas involved, which are then superimposed (intersected) with the pre-registered areas of the crop units. Finally, a last procedure calculates the degree of coverage of the work in the SL on the surface of the individual crop units identified in the previous point.

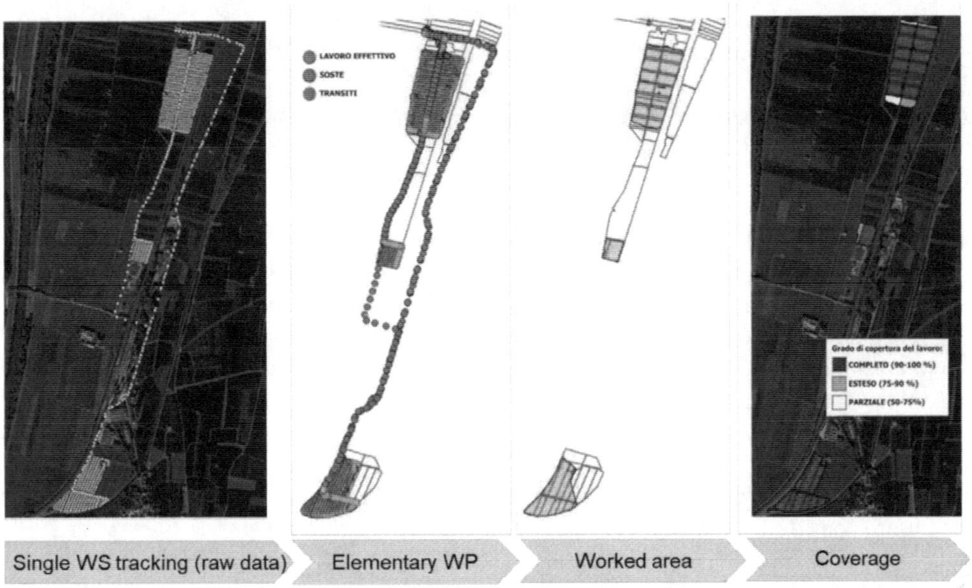

Figure 6. Example of automatic detection of the worked areas, based on tractor's positioning data.

The behavior of the automation system (following the scheme shown in Figure 7) of operational monitoring was tested comparing elementary time recorded by the system and by hand. No statistically significant difference between manual and automatic measures are observed, with a strong correlation for all the elementary times with the exception of maneuvers. This can be due to a not proper working phase assessment done by the algorithm and a stronger work on field speed and pattern complexity have to be done, taking into account also GNSS drift.

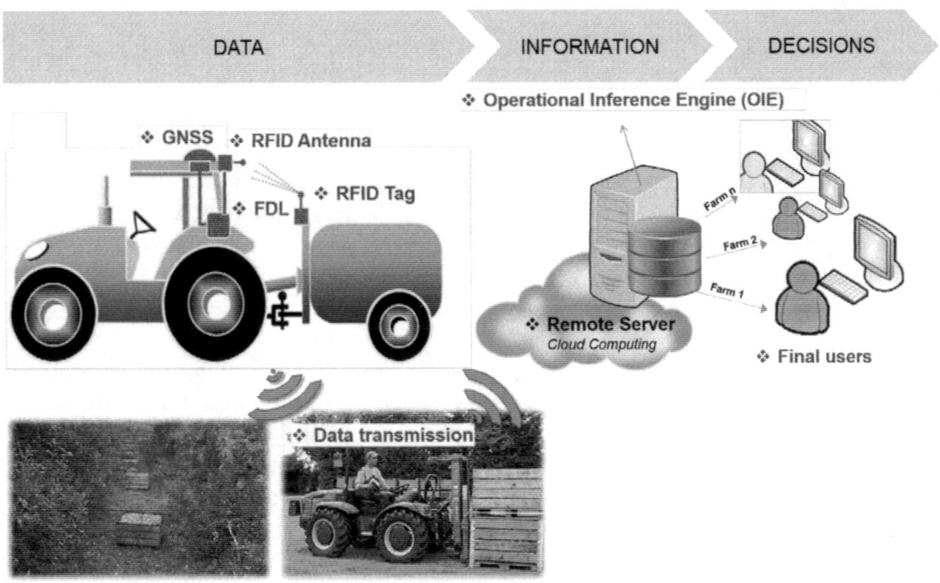

Figure 7. Schema of the operational monitoring system in orchards.

IOP Conf. Series: Earth and Environmental Science **275** (2019) 012013 doi:10.1088/1755-1315/275/1/012013

2.2. Forestry operational monitoring

On its turn, the forestry operational monitoring has mainly focused on manual tree felling operations with chainsaw [5, 6], with the objectives of: 1) identifying the points of felling; 2) providing an estimate of the volumes of the cut plants (through algorithms that estimate the effective cutting times, which are proportional to the cutting section of the trees themselves); 3) estimating the worktimes, including the successive stages for completing the work on the felled trees (declaiming and bucking).

For the vibration assessment a chainsaw was equipped with a tri-axles Wi-Fi accelerometer (10 Hz). The device was fixed thanks to screws at the cover of the air filter (Figure 8). Vibrations and position were measured. During the trial 30 trees were felled, 3 for each diametrical class, for each one the felling time, the stump diameter and the DBH (diameter at 1.30 m from the ground) were manually surveyed and measured. 20 of these parameters were used to develop two mathematical models to calculate the stump diameter and DBH respectively. The amplitude that detect the felling threshold is acceleration > 1.4 ms^{-2}, identified empirically. Through visual assessment each vibration sample was divided into two components the felling and the stem processing (Figure 9). For the research aim only the vibration recorded during felling was taken into account. The effective cut is characterized by having the highest in amplitude (acceleration > 1.4 ms^{-2}). The sum of all these events determine the amount of time spent for the felling.

Figure 8. Measurement equipment on chainsaw.

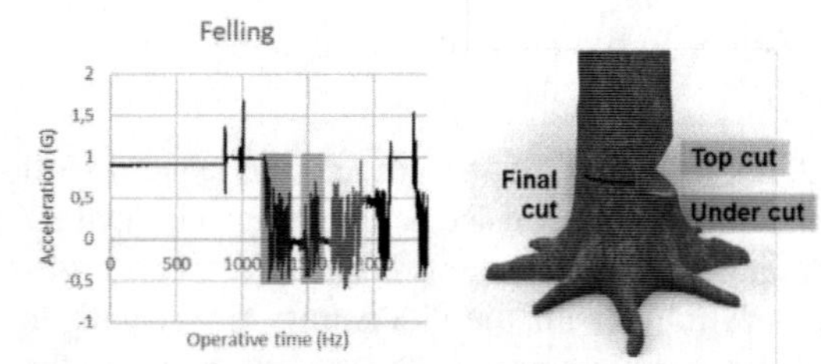

Figure 9. Vibration analysis for productivity evaluation from felling operations.

During the post processing, for each tree the collected dataset was manually split and synchronized with the vibrations dataset. The first GPS and vibration record collected at the same time corresponds at the beginning of the felling operation. Since during the felling the operator is very close to the tree, the points collected inside a radius of 1.5 m, far from the first point, were considered as felling. After that the coordinate of the stamp was calculated averaging the coordinated of the point cloud considered as felling. For both models a satisfactory correlation value was achieved confirming the strong relationship between the variables considered. A very good correlation was obtained also for the volume estimation.

3. Conclusions

The paper presents some applications of farm and forestry operational monitoring based on information systems, considering the whole cycle from data to information ready to use in decision processes. The proposed solution has the goal to summarize useful information in a fully autonomous way. The purpose is achieved through: 1) providing the equipment and the machines in the field of instruments that collects data autonomously without the intervention of the operators and 2) extracting the desired information from the data through suitable algorithms, also in this case without take charge of the end users of data processing tasks.

The first tests carried out to check the reliability of the inference engines highlighting the reliability of the proposed systems to monitor and interpret the operations in an automatic way. Operational Monitoring is fundamental for future information systems supporting Precision Agriculture and Forestry applications and its development must be further enhanced according to a fully automated approach. These solutions will facilitate and enable new generations of farm information systems to be specifically developed for applications on agro-environmental enterprises.

Operational monitoring generates the so called big data, that should be approached in the correct way, then throughout the given cycle data-information-decision. To do this very specific and very different skills are needed: application context to set goals and contextualize information, analytical to extract information from data, management of collection and storage of data, and so on. Furthermore, these are always highly contextualized and personalized processes. It is difficult for a single farm or forestry enterprise to deal with all these necessary skills, and those could represent a barrier to operational monitoring adoption. Probably the best way to proceed is through the creation of centres of expertise, in which all technical and technological parts are entrusted to experts of the individual disciplines and are transparent to end users, who will use them in the form of services.

References

[1] Fountas S, Carli G, Sørensen CG, Tsiropoulos Z, Cavalaris C, Vatsanidou A, Liakos B, Canavari M, Wiebensohn J and Tisserye B 2015 Farm management information systems: Current situation and future perspectives *Computers and Electronics in Agriculture* **115** pp 40–50

[2] Sørensen C G, Pesonen L, Bochtis D D, Vougioukas S G and Suomi P 2011 Functional requirements for a future farm management information system *Computers and Electronics in Agriculture* **76** pp 266–276

[3] Mazzetto F, Gallo R, Importuni P, Petrera S and Sacco P 2017 Automatic filling of field activities register, from challenge into reality *Chemical Engineering Transactions* **58** pp 667–672

[4] Mazzetto F, Riedl M and Sacco P 2016 Sistemi informativi aziendali ed agricoltura di precision. *Agricoltura di precisione. Metodi e tecnologie per migliorare l'efficienza e la sostenibilità dei sistemi colturali* New Business Media

[5] Gallo R, Grigolato S, Cavalli R, and Mazzetto F 2013 GNSS-Based Operational Monitoring Devices for Forest Logging Operation Chains *Journal of Agricultural Engineering* **44** pp 140–144

[6] Gallo R, Nalli F, Cortese L, Knollseisen D, Noggler W, and Mazzetto F 2017 Chainsaw Vibrations, a Useful Parameter for the Automatic Tree Volume Estimations and Production Assessment of Felling Operations *Chemical Engineering Transactions* **58** pp 655–660

IOP Conf. Series: Earth and Environmental Science **275** (2019) 012014 doi:10.1088/1755-1315/275/1/012014

Microsensors for the characterization of the insect gut environment

E Crotti[1]

[1]Department of Food, Environmental and Nutritional Sciences (DeFENS), University of Milan, 20133 Milan, Italy

elena.crotti@unimi.it

Abstract. Microorganisms establish a wide range of symbiotic associations with their insect host, encompassing mutualistic, parasitic and commensal relationships. Insect gut microbiota is involved in many essential aspects of the host biology, physiology and behaviour. Factors such as pH, oxygen availability, retention time of the food and host innate immune system are important determinants of the gut symbionts' colonization. Deciphering the interactions that take place and shape the symbiosis is an intriguing topic that cannot exclude the characterization of the physicochemical conditions occurring in the insect gut niche. Microsensors and microelectrodes can be usefully employed to characterize the physicochemical conditions occurring in the insect intestine, allowing to gain a better knowledge of the gut microenvironment in which the gut microbiota thrives.

1. Introduction

Microbial symbionts are ubiquitous in animals, including mammals and insects, where they contribute to the host health. Particularly, insects had an extraordinary evolutionary success that, partly due to their microbial symbionts, led to their adaptation to very diverse terrestrial and aquatic habitats, including those niches with nutritionally imbalanced or limited regimes e.g. wood, phloem sap or blood [1].

Insect symbionts are classified in different categories, according to the host dependence and the evolutionary age of the association [2]. Primary or obligate symbionts are essential for the host survival; they are generally involved in the nutritional supplementation of the insect and commonly associated with hosts that rely on nutrient-deficient diets. They are vertically transmitted and housed in specialized cells. They typically own a reduced genome and show a congruent evolution with their hosts [2]. Secondary or facultative symbionts are not essential for the host survival, but contribute to the host fitness, for instance by increasing the host resistance against environmental stressors [2]. Within this category, a particular group is represented by reproductive manipulators, the best-known example of which is the Alpha-Proteobacterium *Wolbachia*, responsible for several host reproductive alterations [3]. Finally, gut symbionts are inhabitants of the digestive systems, usually acquired with the diet and subjected to the selection of the intestine physicochemical and biological conditions [4, 5]. Evidences are revealing that they can critically affect the host physiology and biology, influencing larval development [6, 7], host weight gain [8], immune system [9] or mate choice [10]. In order to obtain a better knowledge of the microenvironment in which the microbiota thrives and where the host-microbe interactions that shape the symbiosis take place, a detailed characterization of the physicochemical conditions occurring in the peculiar niche represented by the insect gut is required.

IOP Conf. Series: Earth and Environmental Science **275** (2019) 012014 doi:10.1088/1755-1315/275/1/012014

2. Factors shaping the microbial colonization of the insect gut

The insect gut is a challenging niche for the microbial inhabitants. Factors such as pH, oxygen availability, retention time of the food and host innate immune system are important determinants of the gut symbionts' colonization [4]. For instance, the gut pH is finely regulated and can vary from alkaline (e.g. midguts of lepidopteran larvae), to acidic (e.g. crops of mosquitoes of bees) or neutral values according to the considered compartments. Variations of pH can be also found within a single compartmentalized gut, as in the case of some soil-feeding species with gut pH ranging from 5 to >12 [11, 12]. By the release of fermentation products, gut symbionts can also affect the gut pH and, ultimately, the microbial colonization [4]. Often indicated as a factor that can shape the gut microbiota and drive the microbial colonization, oxygen has been experimentally measured in few examples [8, 11, 13-17]. Oxygen is delivered to the insect gut cells (and microbiota) by the respiratory system, which is based on cuticle-lined tracheae organized in a dense branched apparatus [18]. Considering the restrictions imposed by the diffusive transport of oxygen through epithelia, it has been estimated that the insect gut, in spite of the efficient oxygen consumption by the epithelial cells (in a fraction of a millimeter), is oxic or microoxic [19]. Hence, the insect gut anaerobic niches result from the high microbial oxygen-consuming activity or from particular chemical processes, such as the one occurring in the highly alkaline P1 segment of soil-feeding termites [8, 11, 13-17, 19].

Insect gut has been also indicated as an unstable habitat for microorganisms considering the development-related modifications of the digestive system. Before the adult stage, insects molt several times eliminating, with the foregut and hindgut exoskeletal layers, also the attached microbial populations. Furthermore, the midgut secretes the peritrophic membrane (to surround the bolus) and continually sheds it, discarding in this way the associated microorganisms that generally do not cross it. Finally, in holometabolous insects the metamorphosis results in the remodeling of the gut and other organs with consequences for the microbial partners attached to the gut mucosa [4].

3. Microsensors and microelectrodes to measure physicochemical parameters

Microsensors and microelectrodes have been used to characterize the physicochemical conditions, e.g. pH, oxygen and hydrogen concentrations and redox potential, occurring in the different compartments of the termite gut (e.g. *Nasutitermes corniger* or *Thoracotermes macrothorax*) [11, 16, 20]. For instance, oxygen has been profiled along the gut, evidencing strong and peculiar variations of the oxygenation conditions and highlighting that only the dilated gut regions show anoxic condition in the center [11, 16, 20]. The high microbial oxygen-reducing activity throughout the gut is responsible for the maintenance of anoxic conditions in the gut center [20].

Recently, hypoxic/anoxic gut compartments have been also found in honeybees [8] and mosquitoes [17] (Figure 1). In both cases the use of germ-free insects allowed the detection of higher amounts of oxygen in the germ-free intestines rather than in the ones dissected from conventional insects [8, 17]. Authors have also demonstrated that the hypoxic conditions created by the gut microbiota are essential for the host development [17]. Besides microsensors and microelectrodes, other solutions could be adopted to characterize the physicochemical parameters occurring in the insect intestine. For instance, proper kits have been used to reveal hypoxic conditions of the mosquito gut sections [17] or specific dyes have been administered to the insects (through diet) to detect the pH values of the different *Drosophila* gut compartments [21].

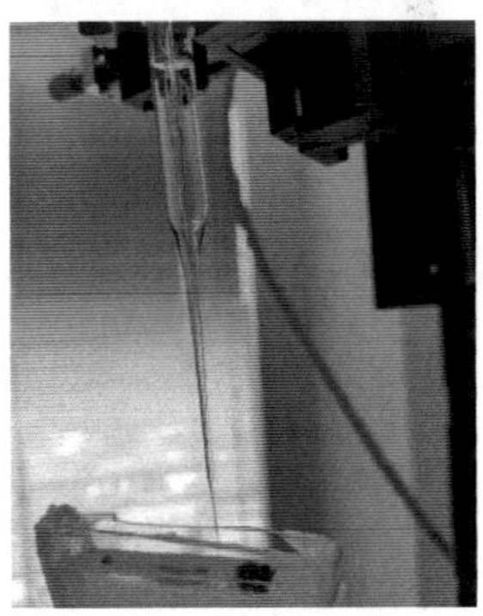

Figure 1. Example of a honeybee gut embedded in agarose gel for oxygen measurement. Courtesy of Matteo Callegari, King Abdullah University of Science and Technology, Saudi Arabia.

4. Conclusions

The composition and structure of the insect gut microbiota is influenced by many factors, such as host diet, genetics and developmental stage. In the last years researchers have devoted many efforts to elucidate their impact on the variation of the host gut microbiota [22-23]. Understanding the drivers that shape the microbial diversity in the insect gut is pivotal to comprehend the insect-microbe interactions. Measurements of the gut physicochemical conditions, as well as the gut metabolite characterization have to be paralleled with the gut microbiota investigations to disentangle the host-microbe interactions.

References

[1] Moran NA 2007 Symbiosis as an adaptive process and source of phenotypic complexity. *Proc. Natl. Acad. Sci. U.S.A.* **104** 8627–33

[2] Dale C and Moran N 2006 Molecular interactions between bacterial symbionts and their hosts. *Cell* **126** 453–65

[3] Werren JH, Baldo L and Clark ME 2008 *Wolbachia*: master manipulators of invertebrate biology. *Nat. Rev. Microbiol.* **6** 741–51

[4] Engel P and Moran NA 2013 The gut microbiota of insects - diversity in structure and function. *FEMS Microbiol. Rev.* **37** 699–735

[5] Prosdocimi E, Gonella E, Mapelli F, Borin S and Crotti E 2015 Microbial ecology-based methods to characterize the bacterial communities of non-model insects. *J Microbiol Methods* **119** 110–25

[6] Shin SC, Kim SH, You H, Kim B, Kim AC, Lee KA, Yoon JH, Ryu JH and Lee WJ 2011 *Drosophila* microbiome modulates host developmental and metabolic homeostasis via insulin signaling. Science **334** 670–4

[7] Chouaia B *et al.* 2012 Delayed larval development in *Anopheles* mosquitoes deprived of *Asaia* bacterial symbionts. *BMC Microbiol.* **12** S2

[8] Zheng H, Powell JE, Steele MI, Dietrich C and Moran NA 2017 Honeybee gut microbiota promotes host weight gain via bacterial metabolism and hormonal signaling. *Proc. Natl. Acad. Sci U.S.A.* **114** 4775–80

[9] Ryu JH, Kim SH, Lee HY, Bai JY, Nam YD, Bae JW, Lee DG, Shin SC, Ha EM and Lee WJ 2008. Innate immune homeostasis by the homeobox gene caudal and commensal-gut mutualism in *Drosophila*. *Science* **319** 777–82

[10] Sharon G, Segal D, Ringo JM, Hefetz A, Zilber-Rosenberg I and Rosenberg E 2010 Commensal bacteria play a role in mating preference of *Drosophila melanogaster*. *Proc. Natl. Acad. Sci U.S.A.* **107** 20051–6

[11] Köhler T, Dietrich C, Scheffrahn RH and Brune A 2012 High-resolution analysis of gut environment and bacterial microbiota reveals functional compartmentation of the gut in wood-feeding higher termites (*Nasutitermes* spp.). *Appl. Environ. Microbiol.* **78** 4691–701

[12] Brune A and Ohkuma M 2010 *Biology of Termites: A Modern Synthesis* eds Bignell DE, Roisin Y & Lo N (Springer, Dordrecht) pp. 439–75

[13] Brune A, Emerson D and Breznak JA 1995 The termite gut microflora as an oxygen sink: microelectrode determination of oxygen and pH gradients in guts of lower and higher termites. *Appl. Environ. Microbiol.* **61** 2681–7

[14] Ceja-Navarro JA, Nguyen NH, Karaoz U, Gross SR, Herman DJ, Andersen GL, Bruns TD, Pett-Ridge J, Blackwell M and Brodie EL 2014 Compartmentalized microbial composition, oxygen gradients and nitrogen fixation in the gut of *Odontotaenius disjunctus*. *ISME J* **8** 6–18

[15] Šustr V, Stingl U and Brune A 2014 Microprofiles of oxygen, redox potential, and pH, and microbial fermentation products in the highly alkaline gut of the saprophagous larva of *Penthetria holosericea* (Diptera: Bibionidae). *J Insect Physiol* **67** 64–9

[16] Brune A and Friedrich M 2000 Microecology of the termite gut: structure and function on a microscale. *Curr. Opi. Microbiol* **3** 263–9

[17] Coon KL, Valzania L, McKinney DA, Vogel KJ, Brown MR and Strand AR 2017 Bacteria-

mediated hypoxia functions as a signal for mosquito development. *Proc. Natl. Acad. Sci U.S.A.* **114** E5362–9
[18] Hetz SK and Bradley TJ 2004 Insects breathe discontinuously to avoid oxygen toxicity. *Nature* **433** 516-9
[19] Brune A, Frenzel A and Cypion H 2000 Life at the oxic-anoxic interface: microbial activities and adaptations. *FEMS Microbiol. Rev.* **24** 691–710
[20] Brune A 1998 Termite guts: the world's smallest bioreactors. *Trends Biotechnol.* **16** 16–21
[21] Shanbhag S and Tripathi S 2009 Epithelial ultrastructure and cellular mechanisms of acid and base transport in the *Drosophila* midgut. *J. Exp. Biol.* **212** 1731–44
[22] Montagna M *et al.* 2015 Effects of the diet on the microbiota of the red palm weevil (Coleoptera: Dryophthoridae). *PLoS One* **10** e0117439
[23] Vacchini V *et al.* 2017 Bacterial diversity shift determined by different diets in the gut of the spotted wing fly *Drosophila suzukii* is primarily reflected on acetic acid bacteria. *Environ. Microbiol. Rep.* **9** 91–103

IOP Conf. Series: Earth and Environmental Science **275** (2019) 012015 doi:10.1088/1755-1315/275/1/012015

A comparison of low-cost techniques for three-dimensional animal body measurement in livestock buildings

A Pezzuolo[1,*], D Giora[1], H Guo[2], Q Ma[2], S Guercini[1] and F Marinello[1]

[1] Department of Land, Environment, Agriculture and Forestry, University of Padova, 35020 Legnaro (PD), Italy.
[2] College of Land Science and Technology, China Agricultural University, Beijing 100083, China.

E-mail: andrea.pezzuolo@unipd.it

Abstract. Data about health and development of animals are still now mostly collected through manual measurements or visual observations but these kinds of methods of collecting data are causes of several problems. Alternatively, optical sensing techniques can be implemented in order to overcome limitations arising from manual contact measurements. The present research discusses metrological analysis of Structure from motion (SfM) photogrammetry approach, low-cost LiDAR scanning and Microsoft Kinect v1 depth camera to three-dimensional animal body measurement, with specific reference to pigs. Analyses were carried out on fiberglass model to get rid of animal movements. Scans were captured based on a segmented approach, where different portion of the body have been imaged during different frames acquisition tasks. The obtained results demonstrate the high potential of 3D Kinect. LiDAR show a higher RMS value respect to Kinect and SfM most probably due to the collection approach based on single profiles rather than on surfaces. Anyway, the RMS of relative noise ranges between 0.7 and 4 mm, showing a high accuracy of reconstructions even for the others techniques.

1. Introduction

Frequent monitoring of animals' body condition is helpful in order to allow for early recognition of health anomalies and consequently decrease the amount of complications related to animal diseases or other stress factors [1-4]. However, such approach is expensive in terms of labor and may be stressful both for the animals and stockman [5].

To overcome such problems, optical methods have been proposed in the last years [6] as a fast non-contact and approach for animal body analysis [7]. Thus optical techniques permit reduction of stress typically induced by manual measurements and at the same time allows fast multi-parameters analysis [8].

Many researches have investigated on three-dimensional reconstruction methods based on optical systems [9-12]. Different kind of optical sensor have been used for agricultural and livestock applications, like for example 2D cameras, but there is a growing interest for 3D sensors, like TOF (Time of Flight) or CTS (Consumer Triangulation Sensor) systems [13].

Some works already mentioned application of the Microsoft Kinect sensor in agriculture [14] and livestock applications [15]. Maki reported quantification of body condition scoring after application of Kinect sensor on cows [16]; McPhee reported results on rump fat and muscle score from low and high muscled Angus cattle [17], while Kongsro reported weight estimation results from two different pig breeds (Landrace and Duroc) in the 20-140 kg interval based on volume information [18].
An alternative approach involves the use of Light Detection and Ranging (LiDAR). LiDAR systems can measure the distance between the sensor and the objects around it very quickly, enabling the construction of three-dimensional point clouds [19]. Through the application of appropriate algorithms, these point clouds can be used to digitally reconstruct and describe the structure of areas [20] or objects of interest with high levels of precision [21-22].
Another promising application of 3D reconstruction that could be implemented in agriculture and livestock's applications is Structure from Motion (SfM). Based on photogrammetry, SfM technique consists in taking photos of the object from all possible angulation and points of view all around the object [23]. Several studies had used this technique in order to collect data about plant's phenology [24], soil topography and roughness [25], but SfM technique could be also implemented for livestock applications [26].
However, available literature is lacking with regard to the metrological performance: this is limiting very much actual application of the methods, since it is not clear how much 3D data can support or replace manual measurements also with reference to different positions of animal body.
The present study proposes a metrological implementation and analysis of 3 low-cost techniques: Microsoft Kinect v1 depth camera, Structure from motion (SfM) photogrammetry approach and LiDAR sensor for reconstruction of pig body. These techniques were applied to an animal fiberglass reference for quantifying the performances of such approach in terms of range, resolution, noise and process time analysis

2. Materials and Methods

2.1. Fiberglass reference

The experimental part of the research consisted in a preliminary validation of the 3 low-cost techniques. To this end, a fiberglass reference artefact resembling the actual shape, posture, colour and dimensions of a real pig was implemented. As typically done in the case of optical instrumentation [6] diffused light condition was used, with an average illuminance of about 400 lux.

2.2. Structure from motion and data processing

SfM is a method based on the estimation of the motions of a camera to allow reconstruction of three-dimensional point-clouds, through the following steps: (*i*) image features detection and description, (*ii*) feature descriptor matching between image pairs, (*iii*) robust pairwise geometry estimation, and (*iv*) 3D point triangulation and transformation of the relative camera poses to a common coordinate frame.
A commercial camera was implemented for collection of images needed in pig SfM reconstruction. Specifically, a Nikon D5100 camera was used, featuring a 23.6×15.6 mm CMOS sensor with a 4928×3264 pixels' resolution and a lens with a 35mm focal length.
For animal side reconstruction, data were collected at a distance between the camera and the animal ranging between 0.7 and 1.0 m; a total of 50 frames were taken, from withers to buttock and from the top of the back to ground.
Three-dimensional reconstruction was carried out through commercial software (AgiSoft PhotoScan, version 1.3.1) allowing reconstruction of three-dimensional point clouds through Structure-From-Motion technique.

2.3. Microsoft Kinect v1 and data processing

For the present research, a commercial Kinect depth camera was implemented for collection of images. Microsoft Kinect is based on coordinate functioning of an infrared laser emitter with an infrared camera and an RGB camera. Three-dimensional reconstruction is achieved through a triangulation process: a diffraction grating splits the infrared laser into a given pattern which is projected onto the scene. The infrared camera than collect the pattern and compare it with the projected one: local shifts are put together to generate a disparity map, where larger shift values correspond to positions further from the sensor and, conversely, lower values correspond to positions closer to the sensor.

However, it has been shown that the output of the sensor is not linear [14] as a consequence, specific calibration procedures have to be carried out. To this end, a substitution method was implemented: a set of references surfaces featuring hemispherical elements and resembling cow body portion geometry was produced and measured. Therefore, hemispheres with radii ranging between 50 and 125 mm and between 150 and 250 mm were implemented for the calibration of the head sensor and of the body sensor respectively.

Three-dimensional data from the Kinect sensor were post-processed by means of commercially available software (SPIP™) undergoing the following operations: (*i*) outliers' management, (*ii*) overlapping images (stitching) and (*iii*) surface generation.

2.4. LiDAR and data processing

LiDAR is a remote-sensing technique for the measurement of the distance between the sensor and a target. Manufactured by Slamtec, RPLidar A3 is a 360° 2D laser scanner with a distance range of 0.15m to 25m, sample rate of 16000 p/s with a scan rate of 20Hz. The laser used has a wavelength of 785nm and 3mW power adhering to FDA Class I laser safety regulations. It is connected to the processor via USB. RPLidar A3 was selected due to its affordable price, the environment of operation, scanning range and weight.

Data were post-processed by means of commercially available software (SPIP™) undergoing the following operations: (*i*) outliers filtering, (*ii*) profiles alignment and (*iii*) surface generation.

3. Results and Discussions

3.1. Methods performances

The discussed applications can benefit from relatively low-cost technologies. The cost range of the instruments is between some tens of Euro (Microsoft Kinect v1 or standard 2D cameras) up to a few hundreds of euros (LiDAR); processing and 3D reconstruction can be carried out through free or low-cost software.

Compared with 3D Kinect imaging, it can be noticed how SfM is still limited by relatively long processing time (needed to produce the 3D model using just 2D images), however such performance is constantly increasing thanks to the availability of computer with faster and better performing processors (Tab.1).

Table 1. Comparison of costs and working times of 3D techniques

Technique	Instrumentation costs[1] [EUR]	Scanning time[2] [min/animal]	Processing time [min/animal]	Body-Parameters extraction [min/animal]
Structure from Motion	50-500	0.3-0.5	90-240	45
Kinect v1	80-150	0.5-0.7	30	30
LiDAR	500-2000	5-7	60	30

[1] Not including tripod, frames, computer and analysis software.
[2] Includes only time to collect data or capture images.

Furthermore, the scanning performance of 3D techniques was repeated at different working distances defined on the basis of actual fiberglass reference measurement, in the range 400-2000 mm.

In general, both horizontal and vertical resolutions normally worsen with the distance (Tab. 2). In the case of SfM, such effect can be reduced taking advantage of proper magnification optics, while is more evident in case of fixed zoom or low-resolution cameras. In the context of LiDAR measurement, the resolution in the slow scanning direction is a function of the number of analysed profiles, but can be as low as 5 mm or even less in most of cases. Finally, Kinect v1 can provide the highest resolution at 0.4 m from the target surface, but rapidly decreases as the distance overcomes 1 m.

Table 2. Comparison of typical achievable x-y-z resolutions with sensors placed at different distances from the target

Distance [mm]	Horizontal resolution (x×y×z) [mm]		
	Structure from Motion	Kinect v1	LiDAR
400	3×3×2	0.65×0.65×0.7	0.4×5×2
1000	3×3×2	1.14×1.14×2	0.8×5×3
2000	3×3×2	1.95×1.95×4	1.6×5×4

Noise and linearity were used to get information about the accuracy of the analyses; they were estimated on a flat surface (roughness <50 µm) and were computed respectively as root mean square (RMS) and maximum deviation from the mean plane.
The obtained results (Tab. 3), in term of noise, demonstrate the high potential of 3D Kinect. LiDAR show a higher RMS value respect to Kinect v1 most probably due to the collection approach based on single profiles rather than on surfaces. SfM exhibits a high variability in terms of linearity: this is due to the fact that such performance depends as much on the accuracy of the processing software, which might difficult in case of flat surface without features or discontinuities useful for the reconstruction.
Maximum detectable slope depends on the capacity of the instruments to capture reflected radiations. This is somehow difficult in the case of Kinect v1 and LiDAR sensors, where extreme slopes could deviate laser beams out of the objective. On the contrary, SfM is not suffering from steep slopes in the target surface, since it relies on multiple repositioning of the camera.
Performances for all of the systems are in general fully acceptable, especially in the case of adult animals where extracted body parameters are typically as big as a few tens of centimetres (Fig. 1).

Table 3. Comparison of performances of 3D techniques

	Structure from Motion	Kinect v1	LiDAR
RMS Noise[1] [mm]	1.0-2.5	0.7-1.2	2.1-4.0
Non-linearity[1] [mm]	10-50	3-6	10-15
Maximum detectable slope [°]	90[2]	45	60-80

[1] Measured on a reference flat surface (RMS roughness <50µm)
[2] Unlimited due to camera repositioning

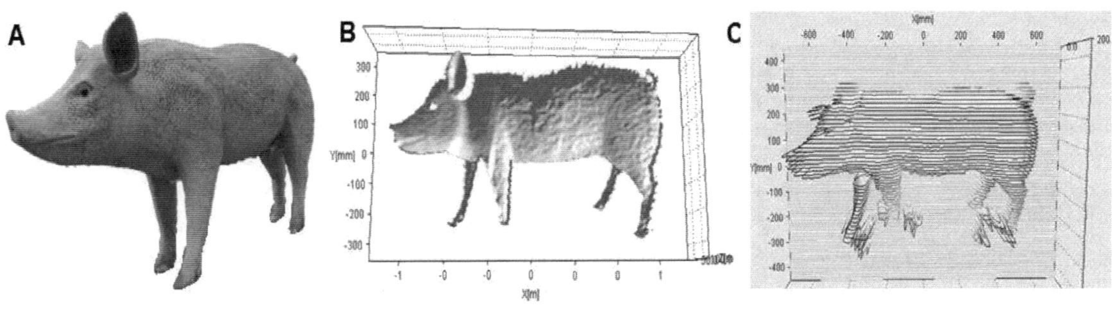

Figure 1. Snapshot of 3D model reconstruction: (A) Structure from motion; (B) Microsoft Kinect v1 and (C) LiDAR.

4. Conclusions

The present research proposes a metrological analysis of 3 low-cost techniques for pig body measurement and non-invasive extraction of quantitative parameters.

Preliminary tests carried out on fiberglass model have shown:

1. Microsoft Kinect v1 is the most cost effective technique, but present application must consider sensitivity to light especially in outdoor environment.
2. Structure from Motion can provide 3D reconstructions with a high resolution, which can be applied for a quantitative extraction of body-parameters, however, the processing time is still relatively long.
3. Low-cost LiDAR is better suited for extraction of specific profiles/sections, requiring precise alignment between animal and sensor source.
4. An application to live animals poses limitations, mainly ascribable to the movements of the animals. Specific approaches will be the object of future developments, oriented to the identification of optimal measurement positions and to the consequent minimization of scanning times.

References

[1] Tsai D and Huang CY 2014 A motion and image analysis method for automatic detection of estrus and mating behaviour in cattle *Comput. Electron. Agric* **104** 25–31

[2] Salau J, Haas J H, Junge W and Thaller G 2017 Automated calculation of udder depth and rear leg angle in Holstein-Friesian cows using a multi-Kinect cow scanning system *Biosyst. Eng* **160** 154–169

[3] Nilsson M, Herlin A, Ardö H, Guzhva O, Åström K and Bergsten C 2015 Development of automatic surveillance of animal behaviour and welfare using image analysis and machine learned segmentation technique *Animal* **9** 1859–1865

[4] Veissier I, Mialon M, Sloth K H 2017 Early modification of the circadian organization of cow activity in relation to disease or estrus *J. Dairy Sci* **100** 3969–3974

[5] Da Borso F, Chiumenti A, Sigura M and Pezzuolo A 2017 Influence of automatic feeding systems on design and management of dairy farms *Journal of Agricultural Engineering* **48** 48-52

[6] Dubbini M, Pezzuolo A, De Giglio M, Gattelli M, Curzio L, Covi D, Yezekyan T and Marinello F 2017. Last generation instrument for agriculture multispectral data collection *CIGR Journal* **19** 158-163

[7] Pezzuolo A, Guarino M, Sartori L and Marinello F 2018 A feasibility study on the use of a structured light depth-camera for three-dimensional body measurements of dairy cows in free-stall barns *Sensors* **18(2)** 673

[8] Nilsson M, Herlin A, Ardö H, Guzhva O, Åström K and Bergsten C 2015 Development of automatic surveillance of animal behaviour and welfare using image analysis and machine learned segmentation technique *Animal* **9** 1859–1865

[9] Wongsriworaphon A, Arnonkijpanich B and Pathumnakul S 2015 An approach based on digital image analysis to estimate the live weights of pigs in farm environments *Comput. Electron. Agric* **115** 26–33

[10] Menesatti P, Costa C, Antonucci F, Steri R, Pallottino F and Catillo G 2014 A low-cost stereovision system to estimate size and weight of live sheep *Comput. Electron. Agric* **103** 33-38

[11] Wang Y, Yang W, Winter P and Walker L 2008 Walk-through weighing of pigs using machine vision and an artificial neural network *Biosyst. Eng* **100** 117–125

[12] Azzaro G, Caccamo M, Ferguson J D, Battiato S, Farinella G M, Guarnera G C, Puglisi G, Petriglieri R and Licitra G 2011 Objective estimation of body condition score by modeling cow body shape from digital images *J. Dairy Sci* **94** 2126–2137

[13] Vázquez-Arellano M, Griepentrog H W, Reiser D and Paraforos D S 2016 3-D imaging systems for agricultural applications - a review *Sensors* **16** 618-626

[14] Vázquez-Arellano M, Reiser D, Paraforos D S, Garrido-Izard M, Griepentrog HW 2018 Leaf Area Estimation of Reconstructed Maize Plants Using a Time-of-Flight Camera Based on Different Scan Directions *Robotics*, **7**, 63

[15] Pezzuolo A, Guarino M, Sartori L, González L A and Marinello F 2018 On-barn pig weight estimation based on body measurements by a Kinect v1 depth camera *Comput. Electron. Agric* **148** 29-36

[16] Maki N, Nakamura S, Takano S and Okada Y 2017 3D Model Generation of Cattle Using Multiple Depth-Maps for ICT Agriculture. In Conference on Complex Intelligent and Software Intensive Systems 768–777

[17] McPhee M J, Walmsley B J, Skinner B, Littler B, Siddell J P, Cafe L M, Alempijevic A. 2017 Live animal assessments of rump fat and muscle score in Angus cows and steers using 3-dimensional imaging *J. Anim. Sci* **95** 1847–1857

[18] Kongsro J 2014 Estimation of pig weight using a Microsoft Kinect prototype imaging system *Comput. Electron. Agric* **109** 32–35

[19] Sanz-Cortiella R, Llorens-Calveras J, Escolà A, Arnó-Satorra J, Ribes-Dasi M, Masip-Vilalta J, Camp F, Gràcia-Aguilá F, Solanelles-Batlle F, Planas-DeMartí S, Pallejà-Cabré T, Palacin-Roca J, Gregorio-Lopez E, Del-Moral-Martínez I and Rosell-Polo J R 2011 Innovative LIDAR 3D Dynamic Measurement System to Estimate Fruit-Tree Leaf Area *Sensors* **11** 5769-5791

[20] Rottensteiner F and Briese C 2002 A new method for building extraction in urban areas from high-resolution LIDAR data *Remote Sensing and Spatial Information Sciences* **34** 295-301

[21] Rosell J R, Llorens J, Sanz R, Arno J, Ribes-Dasi M, Masip J, Escolà A, Camp F, Solanelles F and Gràcia F 2009 Obtaining the three-dimensional structure of tree orchards from remote 2D terrestrial LIDAR scanning *Agr. Forest. Meteorol* **149** 1505-1515

[22] Hosoi F, Nakabayashi K and Omasa K 2011 3-D Modeling of tomato canopies using a high-resolution portable scanning lidar for extracting structural information *Sensors* **11** 2166-2174

[23] Westoby M J, Brasington J, Glasser N F, Hambrey M J, and Reynolds J M 2012 Structure from Motion photogrammetry: A low-cost effective tool for geoscience applications *Geomorphology* **179** 300-314

[24] Jay S, Rabatel G, Hadoux X, Moura D and Gorretta N 2015 In-field crop row phenotyping from 3d modeling performed using structure from motion *Comput. Electron. Agric* **110** 70-77

[25] Javernick L, Brasington J and Caruso B 2014 Modelling the topography of shallow braided rivers using Structure-from-Motion Photogrammetry *Geomorphology* **213** 166–182

[26] Pezzuolo A, Milani V, Zhu D, Guo H, Guercini S and Marinello F 2018 On-Barn Pig Weight Estimation Based on Body Measurements by Structure-from-Motion (SfM) *Sensors* **18** 3603

[27] Savio E, De Chiffre L, Schmitt R, 2007 Metrology of freeform shaped parts *CIRP Annals Manufacturing Technology* **56** 810-835

IOP Conf. Series: Earth and Environmental Science **275** (2019) 012016 doi:10.1088/1755-1315/275/1/012016

Multi-sensor platform for real time measurements of honey bee hive parameters

S. Cecchi[1], **A. Terenzi**[1], **S. Orcioni**[1], **S. Spinsante**[1], **V. Mariani Primiani**[1], **F. Moglie**[1], **S. Ruschioni**[2], **C. Mattei**[2], **P. Riolo**[2], **N. Isidoro**[2]

1 DII - Universita Politecnica delle Marche - Italy
2 D3A - Universita Politecnica delle Marche - Italy

E-mail: s.cecchi@univpm.it

Abstract. Honey bees (*Apis mellifera L.*) are well-known insects that have positive effects on the environment and on the human life. They are so important that the decline of honey bee colonies happened in the last years has triggered an increasing interest for their safeguard. In this context, the proposed work aims at developing a multi-sensor platform to monitor the beehives' conditions in real time, based on the measurement of sounds emitted by the bees, temperature, humidity, CO_2, weight inside the beehive, and weather conditions outside the hive. In this paper, a detailed description of the multi-sensor platform is reported and its application in a real scenario is presented.

1. Introduction

The importance of honey bees is not limited to the production of honey, beeswax, royal jelly, and propolis, because they are at the basis of the plants' pollination, and play a key role in the proliferation of both spontaneous and cultivated flora. Recent years have witnessed an increase in bees' mortality which can result in a loss of pollination services, with remarkably negative ecological and economic impacts that could significantly affect the maintenance of wild plant diversity, the wider ecosystem stability, and the crop production. The causes can be found in illnesses, like the Colony Collapse Disorder (CCD) which is characterized by a sudden disappearance of honey bees from the hive [1, 2]. Many bee scientists agree that the decline of honey bee colonies is the result of multiple stressors, acting independently, in combination, or synergistically to impact honey bees' health [2, 3]. Starting from the above-mentioned context, the necessity of an intensive monitoring activity of honey bees emerges clearly, in order to understand the problems and the causes of bees' mortality.

Monitoring sounds inside the hive is very important, since sound is used by bees to communicate within the colony. Sound generation by a bee takes place through several mechanisms: body movements, wing movements, high-frequency muscle contractions without wing movements, and by pressing the thorax against the substrates or another bee. A strict relation between sound characteristics and some events within the hive, like swarming or the presence of the queen bee, has been already proved [4],[5],[6].

Temperature and humidity are important parameters to measure inside and outside the hive, since they influence the bees' health, the brood, and the productivity of the beehive. Several studies have been carried out on this topic showing that a correct temperature and humidity can significantly decrease the mortality rate in the colony and can increase the honey production.

Carbon dioxide (CO_2) measurement is another important aspect since it is related to the bees' metabolism. In particular, metabolic heating of a bee normally accompanies a change in the respiratory emission of CO_2. As the carbon dioxide within a colony in the hive can reach much higher levels than the normal atmospheric ones, honey bees use fanning and gas exchange events to expel CO_2-rich air, and to keep the CO_2 at an acceptable level [7],[8],[9].

Finally, regularly weighing the hives can provide useful information on the changes over time in the main biological components of honey bee colonies [10], i.e., adult and brood populations and food stores[11],[12]. Weighing honey bee hives often provides precise information on timing and size of events such as swarming and unforeseen hive phenomena.

Starting from the aspects presented above, a multi-sensor platform [13] has been designed and implemented, considering two different modules, in order to have several satellite boards on each hive, and a single master for one colony. The system is capable of acquiring in real time sounds, temperature, humidity, CO_2, and weight inside the beehive, and weather-related data outside the hive. All these parameters are strictly related to the bees' health status and they can be used to analyze and to forecast critical and possibly dangerous situations. The system has been installed in real bee hives and it will be used to collect a complete dataset of the aforementioned parameters.

The paper is organized as follows. Section 2 describes the hardware platform development, considering the selected sensors and their physical deployment. Section 3 details the software platform and the functions developed to evaluate the measured values. The deployment of the multi-sensor platform in a real scenario, and the related issues, are reported in Section 4, showing the behavior of some measured parameters. Finally, conclusions are reported in Section 5.

2. Hardware platform for data acquisition

The hardware platform is composed by two different modules, in order to have a set of satellite boards (called Bee Boards, one for each hive) and a single master board (called Queen Board) for one colony. Figure 1(a) shows a schema of the Bee Board module installed in each hive. The module is capable of recording signals from input sensors, i.e., sound and microclimatic parameters. The sensors are positioned inside the hive in a way that they result as less invasive as possible. This is an important aspect to avoid some issues such as the covering of the sensing devices with propolis by the bees, as they usually do with every foreign body entering the hive [14].

For sound acquisition, two microphones are installed in each hive. MEMS (Micro Electro-Mechanical Systems) microphones have been chosen in order to keep the whole system as hidden and smaller as possible. In particular, the chosen model is the Analog Devices ADMP401 [15], featuring a bandwidth between 100 Hz and 15 kHz. Signals from the microphones are acquired by a 24 bit USB sound card, connected to the Raspberry Pi and providing a sample rate of 32 kHz.

In order to acquire temperature and humidity data, the DHT22 [16] sensor has been chosen, being a low cost and small one-wire digital sensor. Each hive is equipped with three DHT22 sensors, two inside the hive for data acquisition, and one located near to the electronic board for system health monitoring. Regarding temperature, each sensor features a measurement range from $-40\,°C$ to $80\,°C$, with an accuracy of $\pm 0.5\,°C$ and a resolution of $0.1\,°C$. Moreover, relative humidity is measured in the range $[0 \div 100]\%$ Rh with an accuracy of $\pm 2\%$ Rh and a resolution of $\pm 0.1\%$ Rh.

For carbon dioxide, the Telaire TL6615 [17] sensor has been installed. It is realized in Non Dispersive Infrared (NDIR) flow-through technology. The TL6615 can measure in the range $[0 \div 50000]$ ppm with an accuracy of 75 ppm on 10% of reading, generating an analog output

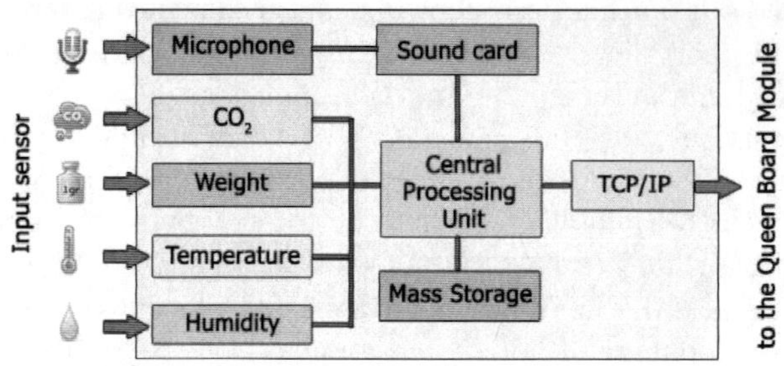

(a)

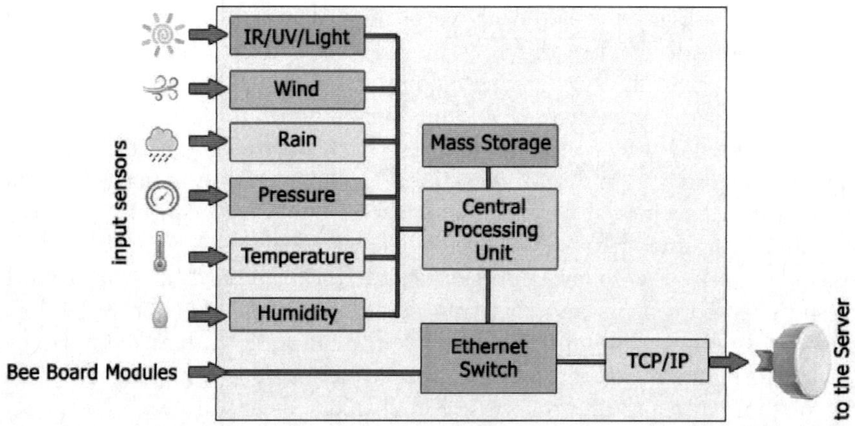

(b)

Figure 1. (a) Bee Board module and (b) Queen Board module .

from 0 V to 4 V, which is acquired by a Texas Instruments ADS115, an I^2C Analog to Digital Converter (ADC), capable of 16 bit resolution and a sample rate of 860 Hz.

Finally, the weight is measured by means of four SparkFun weight sensors. Each sensor is an analog load cell with a flow of 50 kg [18], connected to the other cells in a Wheatstone bridge configuration. The bridge output is acquired by an HX711 [19] 24 bit ADC specific for weighting application.

The embedded data collection platform is managed by the central processing unit, i.e., a Raspberry Pi 3 Model B programmed on purpose, that acquires data generated from each input sensor. Figure 2 shows the hardware housing, consisting of the Raspberry Pi, the sound card, the analog interface to the sensors and the power supply.

Figure 1(b) shows the overall scheme of the master module that collects the data related to weather conditions and manages the data generated from each hive. The module is installed near the colony and communicates with each hive over a TCP/IP connection. The master also connects the entire system to the outside world through a wireless high-speed Internet connection supported by a point-to-point radio link implemented with MikroTik SXTG-5HPnD-SAr2 antennas . This way, a PC server is directly connected to the bee colony, running a software developed ad hoc to manage all the acquired data. The software moves the data from the mass storage of each bee hive into the server, and allows the real-time visualization of the data, through a graphical interface.

Figure 2. Hardware installation outside the bee hive.

3. Software components in the data acquisition platform

The software components necessary for the data acquisition platform have been developed using two programming languages: the Raspberry Pi acquisition code has been written in Pyhton 2.7, while the server tasks for data visualization and data storage have been implemented using LabView 2016. Focusing on the Python script, the code reads data from each sensor every five seconds; sensor data are written in .csv file and sent to the remote server for real-time visualization, over a TCP connection. Moreover, exploiting the functionalities of PyAudio and the Advanced Linux Sound Architecture (ALSA) driver, sounds from the hive are recorded continuously, and saved every ten minutes on .wav files. At the server side, different LabView Virtual Instruments (VIs) have been developed. Figure 3(a) reports a screen shoot of a VI that shows in real-time the data acquired from each bee hive and received through the TCP connection, i.e., temperature and humidity, weight and CO_2 values. Figure 3(a) shows a screen shoot of a VI that acquires data from the weather station i.e., temperature, humidity, atmospheric pressure, UV-light, visible light, infrared light, rain wind intensity and wind direction. Finally, Figure 3(c) shows a screen shoot of a VI that allows a preliminary sound analysis on recorded data, by means of Fast Fourier Transform (FFT). During the data acquisition process, ad-hoc script move data from each Raspberry Pi internal storage to a remote Network Attached Storage where data are saved and available for analysis.

4. A case of study

The system has been deployed in the field, to monitor a colony composed by three hives positioned within the University Campus. Figure 4 shows the installation of the system with a Bee Board module deployed in each hive and the Queen Board module, equipped with the weather station sensors, and the antenna for the radio communication to the remote server. The data recording started in 2017, June 15th: until now, 17 TB of data have been acquired. Particular events with a great relevance for the bees' health have been annotated in the database, such as the death of the queen bee, bee hives diseases, and the presence of the Varroa mite. Figures 5(a) and 5(b) show sample data acquired from the microphone of one hive, over a period of 10 minutes, in time and frequency domain, respectively. As reported in [13, 6], the sound can vary in several situations, e.g., when the queen bee is dead or in a warning condition. Future works will be oriented to the development of an automatic procedure for real-time sound analysis and event detection. Figures 5(c) to 5(f) show the variation of CO_2, weight, humidity and temperature during one day. An evident correlation among all the parameters appears:

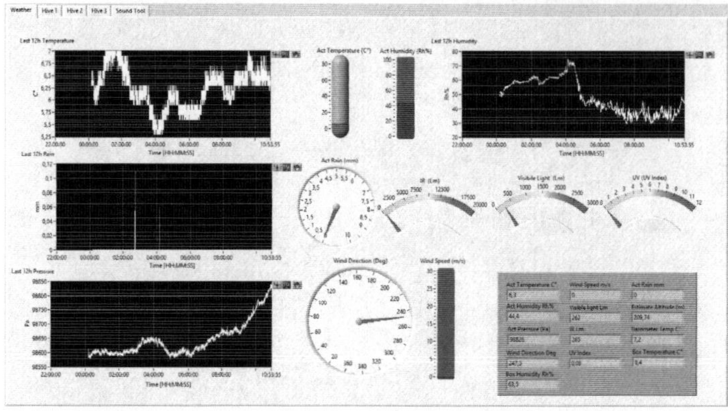

(a)

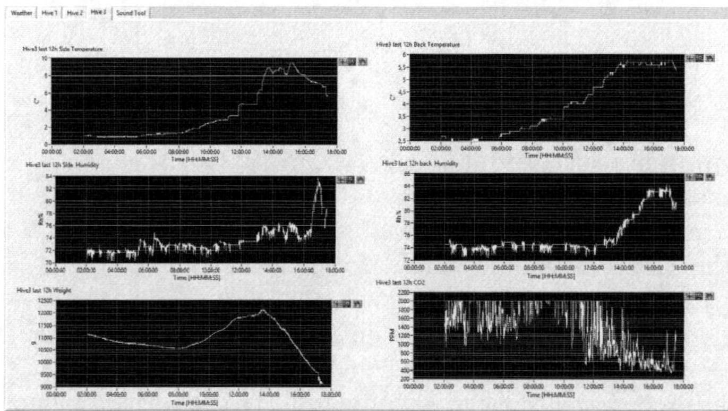

(b)

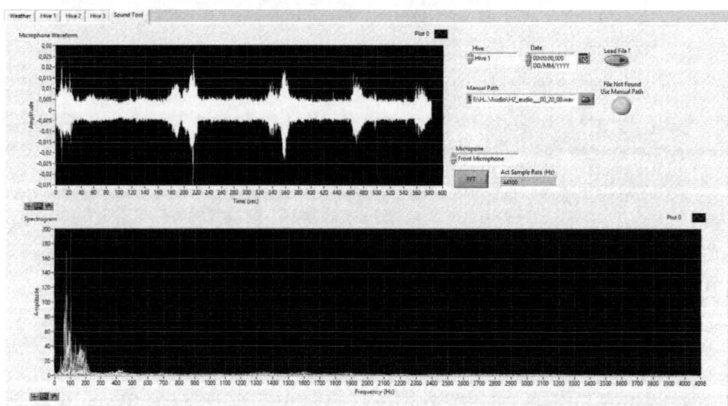

(c)

Figure 3. Graphical user interface of the software developed for real-time monitoring and data acquisition. In particular: (a) visualization of the Queen Board data together with micro climatic parameters, (b) visualization of the data generated by each Bee Board (one virtual screen for each hive), (c) visualization of the sound emitted by a selected hive, in time (top screen) and frequency (bottom screen) domain, in arbitrary units (a.u.).

Figure 4. In the field deployment of the Bee Board modules and the Queen Board module.

during the day most of the bees go outside and there is a reduction of the hive's weight, of the CO_2 and of the temperature, going to the detriment of the humidity. Future works will be oriented to a joint analysis of these parameters, as a function of the sound generated within the bee hives.

5. Conclusions

In this work, a multi-sensor platform capable of real-time monitoring beehives condition has been presented. Such a system has been developed to record the sounds emitted by the bees in the hives, and also other measurable parameters such as temperature, humidity, CO_2, hive weight and weather conditions. A real deployment of the system in the field has been described, joint with an overview of the preliminary results obtained from the measurement campaign that is going on since 2017. Future works will be oriented to the development of tools for automatic data analysis and event detection, aimed at a better understanding of the conditions affecting the bee colony's health status.

References

[1] J. Faucon, L. Mathieu, M. Ribire, A. Martel, P. Drajnudel, S. Zeggane, and et al., "Honey bee winter mortality in france in 1999 and 2000," *Bee World*, vol. 83, pp. 14–23, 2002.

[2] B. Oldroyd, "Whats killing american honey bees?," *PLOS Biol*, vol. 5, 2007.

[3] D. Van Engelsdorp, J. J. Hayes, R. Underwood, and P. J.S., "A survey of honey bee colony losses in the united states, fall 2008 to spring 2009," *J Apic Res.*, vol. 49, pp. 7–14, 2010.

[4] D. Dietlein, "A method for remote monitoring of activity of honeybee colonies by sound analysis," *Journal of Apicultural Research*, vol. 24, no. 2, pp. 176–183, 1985.

[5] S. Ferrari and et al., "Monitoring of swarming sounds in bee hives for prevention of honey loss," in *International Workshop on Smart Sensors in Livestock Monitoring*, Sep. 2006.

[6] A. Qandour, I. Ahmad, D. Habibi, and M. Leppard, "Remote beehive monitoring using acoustic signals," 2014.

[7] G. Nicolas and D. Sillans, "Immediate and latent effects of carbon dioxide on insects," *Annual Review of Entomology*, vol. 34, no. 1, pp. 97–116, 1989.

[8] E. Southwick and R. Moritz, "Social control of air ventilation in colonies of honey bees, apis mellifera," *Journal of Insect Physiology*, vol. 33, no. 9, pp. 623 – 626, 1987.

[9] T. Seeley, "Atmospheric carbon dioxide regulation in honey-bee (apis mellifera) colonies," *Journal of Insect Physiology*, vol. 20, no. 11, pp. 2301 – 2305, 1974.

[10] W. Meikle, N. Holst, G. Mercadier, F. Derouan, and R. James, "Using balances linked to dataloggers to monitor honey bee colonies," *Journal of Apicultural Research*, vol. 45, no. 1, pp. 39–41, 2006.

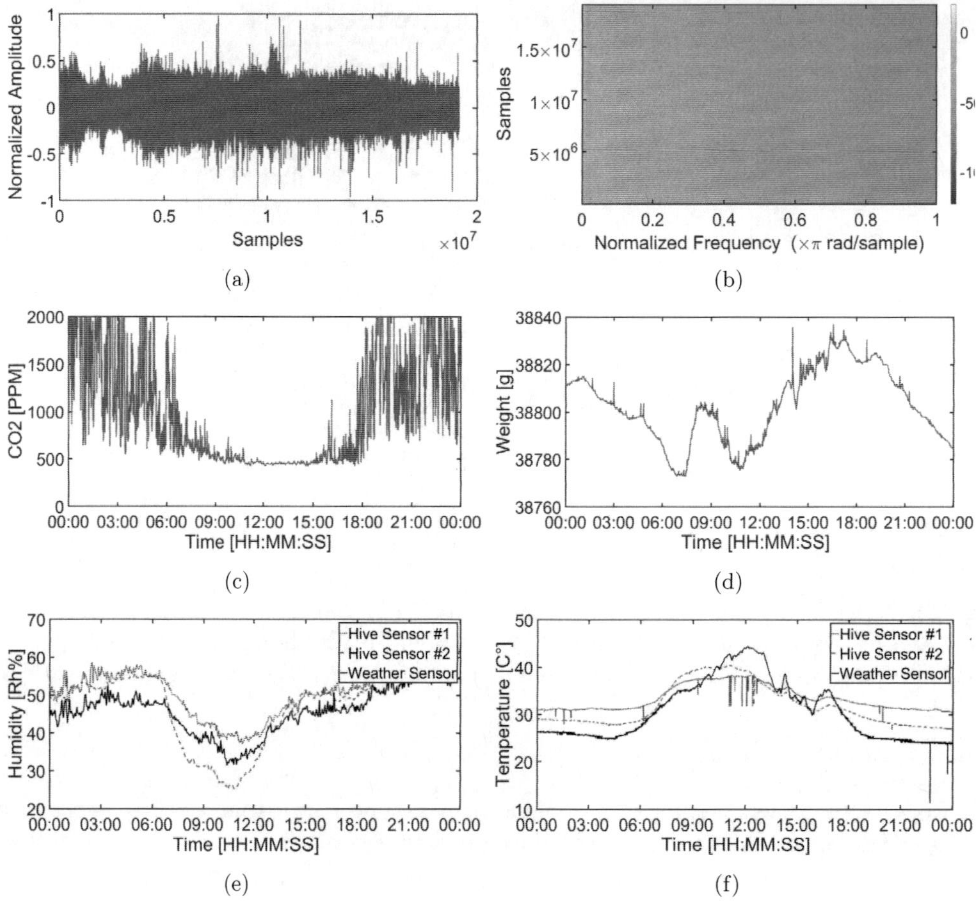

Figure 5. Sample data acquired from the hive sensors: (a) sound in time domain (sampling frequency: 32 kHz), (b) sound in frequency domain, (c) CO_2 inside the hive, (d) hive weight, (e) relative humidity and (f) temperature in the hive.

[11] A. McLellan, "Honeybee colony weight as an index of honey production and nectar flow: A critical evaluation," *Journal of Applied Ecology*, vol. 14, no. 2, pp. 401–408, 1977.

[12] J. R. Harbo, "Worker-bee crowding affects brood production, honey production, and longevity of honey bees (hymenoptera: Apidae)," *Journal of Economic Entomology*, vol. 86, no. 6, pp. 1672–1678, 1993.

[13] S. Cecchi, A. Terenzi, S. Orcioni, P. Riolo, S. Ruschioni, and N. Isidoro, "A preliminary study of sounds emitted by honey bees in a beehive," in *Audio Engineering Society Convention 144*, May 2018.

[14] S. Gil-Lebrero, F. Quiles Latorre, M. Ortiz, V. Snche Ruiz, V. Gamiz, and J.-J. Luna-Rodriguez, "Honey bee colonies remote monitoring system," *Sensors*, vol. 17, p. 55, 01 2017.

[15] "Admp401 mems microphone." https://www.analog.com/media/en/technical-documentation/obsolete-data-sheets/ADMP401.pdf.

[16] "Dht22 temperature humidity temperature sensor." https://www.sparkfun.com/datasheets/Sensors/Temperature/DHT22.pdf.

[17] "Telaire tl6615 co2 sensor." https://www.amphenol-sensors.com/en/telaire/co2/525-co2-sensor-modules/319-t6615.

[18] "Sparkfun 50kg load sensor." https://www.sparkfun.com/datasheets/Sensors/loadsensor.pdf.

[19] "Hx711 weight scale adc." https://www.mouser.com/ds/2/813/hx711_english-1022875.pdf.

IOP Conf. Series: Earth and Environmental Science **275** (2019) 012017 doi:10.1088/1755-1315/275/1/012017

A novel approach for measuring treeline spatial complexity

S. Spinsante[1], L. Montalto[2], M. Garbarino[3], F. Malandra[4], P. J. Weisberg[5], N. Paone[6], L. Scalise[6]

[1]Dept. of Information Engineering, Università Politecnica delle Marche (Italy)
[2]Dept. of Materials, Environmental Sciences and Urban Planning, Università Politecnica delle Marche (Italy)
[3]Dept. of Agricultural, Forestry and Food Sciences, Università degli Studi di Torino (Italy)
[4]Dept. of Agricultural, Food and Environmental Sciences, Università Politecnica delle Marche (Italy)
[5]Dept. of Natural Resources and Environmental Science, University of Nevada, Reno (USA)
[6]Dept. of Industrial Engineering and Mathematical Sciences, Università Politecnica delle Marche (Italy)

E-mail: `s.spinsante@staff.univpm.it`

Abstract. Treelines, defined as *ecotonal zones between closed forest and the uppermost trees*, are particularly sensitive to global changes related to climate and anthropic activities. Different mechanisms of treeline formation can be detected as subtle differences in ecotonal structure, which in turn have important implications for how treelines function and potentially respond to global changes. So, it is of interest to be able to measure in a precise and quantitative way treelines' properties reflecting climate and land use changes. Classical tools adopted to measure treeline spatial patterns are not able to fully understand the limiting factors affecting them. This work presents a novel textural analysis of treeline spatial structure based on the measurement of surface roughness, and applies the corresponding metrics to twenty study areas at both Upper and Lower treelines, where all tree crowns have been mapped at high precision. Preliminary results are promising and motivate future and more extensive evaluations on bigger datasets.

1. Introduction

Treelines are transition zones whose position is widely thought to be temperature sensitive, and potentially responsive to climate warming [1]. For this reason, the dynamics of treelines are studied around the globe with the aim of detecting changes, understanding responses to temperature variation, and evaluating the threats to alpine and arctic biota in response to treeline movement, in high altitude and latitude communities [2]. At a landscape and local scales, treeline position, form (spatial pattern), and dynamics depend on multiple interactions of influencing factors and mechanisms [3], including human activities.

In some regions of the world, the altitudinal limit of the forest also called altitudinal treeline may be above (Upper treeline) or below (Lower treeline) the closed canopy forest. It is assumed that Upper and Lower treelines, like those shown in Figure 1, are conditioned by two different limiting factors: temperature and humidity, respectively [4]. The appearance of treelines, i.e. their spatial pattern, seems to be characterized by grouped trees or microsites in the case of

Figure 1. A visual example of Upper and Lower treelines, mostly affected by temperature (T) and water availability (H_2O), respectively.

Upper treelines, whereas it is influenced by water availability or random effects in the Lower treelines. As a result, treelines' pattern is typically linear or at least regular in the Lower case, whereas it appears to be complex and spread in the Upper case.

In order to objectively analyze the dynamics of treelines, like changes in their shape and pattern over time, it is important to adopt a methodology based on clearly defined parameters, that can be quantitatively measured and compared. Given the fact that, at the authors' best knowledge, such a methodology has not been presented up to now, in this paper we propose to apply typical approach and parameters used in the domain of surface roughness measurements, to the characterization of treelines features and properties. The reference standard in Geometrical Product Specifications (GPS), the ISO 4287:1997 one, that was confirmed in 2015 [5], prescribes the profile method for surface texture characterization, and provides terms, definitions and surface texture parameters. We refer to this standard in order to identify the metrics that can be reasonably applied to characterize treelines.

The paper is organized as follows: Section 2 introduces the metrics used in surface roughness measurements, and specifies how we applied them to treelines' data. Section 3 describes the experiments performed and discusses the results obtained. Finally, Section 4 draws the main conclusions of this work and highlights future developments.

2. Surface Roughness Metrics and Their Application to Treelines

The application of surface roughness measurements to treelines is possible because, in practice, measured surface texture data is not continuous but takes discrete values, similarly to the nature of the treeline data we have to process. In fact, treelines are given in the form of sets of points, each one corresponding to the position of a tree in a 2D view of a geographic area, representing a mountain slope. Each point is identified by its horizontal and vertical coordinates (x,y), within a common reference system. Figure 2(a) shows a graphical representation of a generic set of treeline data: each point in the graph identifies a tree. Figure 2(b) shows the same set of data, over which three trend curves have been fitted. Curve labelled as no. 1 corresponds to a degree-6 polynomial fitting the highest vertical coordinates of trees falling within the same 5 m - wide interval along the horizontal axis, whereas curve no. 2, that is very similar, shows a degree-6

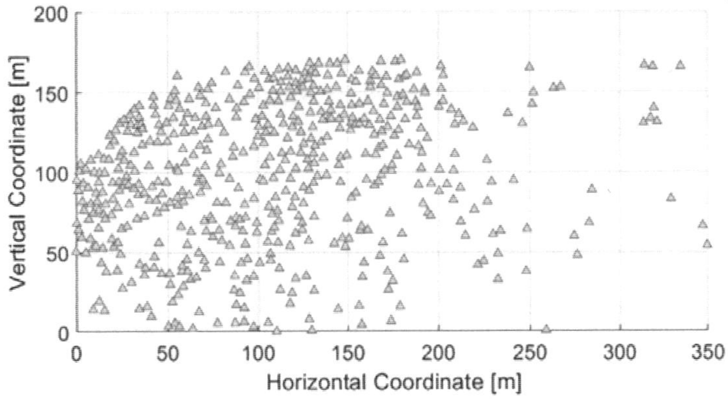

(a)

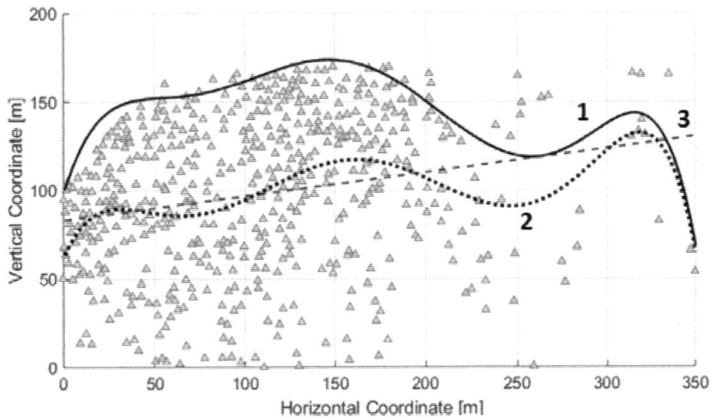

(b)

Figure 2. Graphical representation of a generic set of treeline data: (a) trees' coordinates, (b) trend curves.

polynomial fitting the single trees' vertical coordinates. Finally, curve no. 3 is the set regression line, fitted by the method of least squares.

All the three curves in Figure 2(b) are obtained by interpolating the available trees' coordinates; however, this gives rise to synthetic values that do not correspond to actual trees in the observed area. In order to avoid the need of generating interpolated values, we apply the following surface texture parameters, that can be computed over discrete data:

- Average Line (AL), the dashed one in Figure 2(b): it is the line fitted by the method of least squares, computed over the primary roughness profile points. In its definition, the AL corresponds to the previously discussed curve no. 3 in Figure 2(b);
- R_a: it is the most commonly used and recognized parameter to evaluate a surface roughness profile, i.e. the arithmetic average absolute distance of the roughness profile points from the AL:

$$R_a = \frac{1}{l}\sum_{i=1}^{N}|d_i| \tag{1}$$

where l is the length of the AL, N is the number of points in the profile, and d_i is the Euclidean distance of each point from the AL, identified by the sample dashed arrows shown

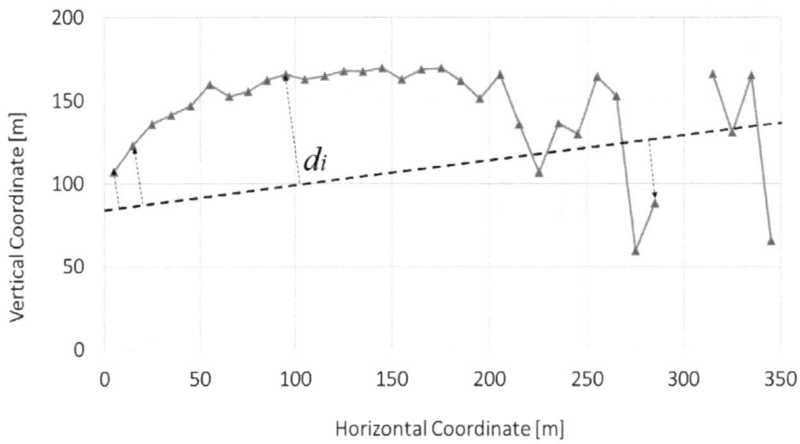

Figure 3. Graphical representation of R_a computation from trees' coordinates.

in Figure 3. The last one shows the line extracted from the trees' coordinates shown in Figure 2(a) by taking the highest tree in vertical coordinate every 5 m along the horizontal direction (x-axis). This line could be called "tree species line" using the ecological meaning of a line passing through the trees that live at the top of the slope;

- R_q: it is the RMS corresponding to R_a:

$$R_q = \sqrt{\frac{1}{l}\sum_{i=1}^{N} d_i^2} \tag{2}$$

- R_{sk} (skewness): it measures the symmetry of the roughness profile with respect to the AL, and can be used to highlight the different symmetry of roughness profiles exhibiting similar R_a and R_q:

$$R_{sk} = \frac{1}{R_q^3}\left[\frac{1}{l}\sum_{i=1}^{N} d_i^3\right] \tag{3}$$

- R_{ku} (kurtosis): it measures the profile acuity:

$$R_{ku} = \frac{1}{R_q^4}\left[\frac{1}{l}\sum_{i=1}^{N} d_i^4\right] \tag{4}$$

Additionally, parameters R_v^*, R_p^*, and R_t are considered, providing, respectively: the maximum distance from the AL of the profile point located under it; the maximum distance from the AL of the profile point located over it; and the corresponding distance between these two points.

3. Experiments and Results

IOP Conf. Series: Earth and Environmental Science **275** (2019) 012017 doi:10.1088/1755-1315/275/1/012017

Table 1. Roughness parameters evaluated over 10 UP treelines, with average (AVG) values in the last line.

Treeline	R_a	$R_q[m^{-1}]$	$R_{v^*}[m]$	$R_{p^*}[m]$	$R_t[m]$	$R_{sk}[m^{-1}]$	$R_{ku}[m^{-2}]$
UP1	37.04	44.86	96.26	82.66	178.92	-0.50	2.42
UP2	55.94	60.16	75.04	100.26	175.30	0.06	1.37
UP3	35.51	43.32	87.00	51.96	138.96	-0.90	2.38
UP4	26.94	34.45	78.01	77.40	155.41	0.05	2.77
UP5	32.04	38.14	69.38	81.19	150.57	0.07	2.26
UP6	26.34	31.67	83.03	51.39	134.43	-0.43	2.28
UP7	24.70	31.95	87.84	57.34	145.18	-0.80	3.53
UP8	33.36	38.71	70.14	68.46	138.60	0.02	1.87
UP9	52.00	56.76	106.00	68.38	174.39	-0.42	1.57
UP10	21.25	27.14	95.92	47.69	143.61	-0.79	4.33
AVG	34.51	40.72	84.86	68.67	153.54	-0.36	2.48

Table 2. Roughness parameters evaluated over 10 LOW treelines, with average (AVG) values in the last line.

Treeline	R_a	$R_q[m^{-1}]$	$R_{v^*}[m]$	$R_{p^*}[m]$	$R_t[m]$	$R_{sk}[m^{-1}]$	$R_{ku}[m^{-2}]$
LOW1	30.31	35.36	87.83	47.76	135.6	-0.65	2.41
LOW2	14.54	17.84	36.19	44.67	80.87	0.01	2.75
LOW3	30.33	37.55	47.81	99.23	147.04	0.89	2.85
LOW4	34.78	42.98	96.57	77.42	173.98	-0.59	2.61
LOW5	35.81	42.74	106.53	65.80	172.32	-0.46	2.65
LOW6	32.93	39.47	74.45	78.91	153.36	0.39	2.17
LOW7	24.11	30.19	107.97	50.16	158.13	-0.75	4.05
LOW8	26.02	31.61	70.89	55.00	125.89	-0.29	2.26
LOW9	41.21	48.20	102.17	66.73	168.89	-0.59	2.14
LOW10	40.01	48.18	119.35	59.68	179.03	-0.6	2.31
AVG	31.00	37.41	84.98	64.54	149.51	-0.26	2.62

The roughness parameters introduced in the previous section have been computed on 10 datasets representing transect plots of mapped trees (named UP1 to UP10) describing the Upper treeline, that is basically influenced by temperature variations, and on other 10 datasets (named LOW1 to LOW10) providing the Lower treeline, that is usually affected by water availability.

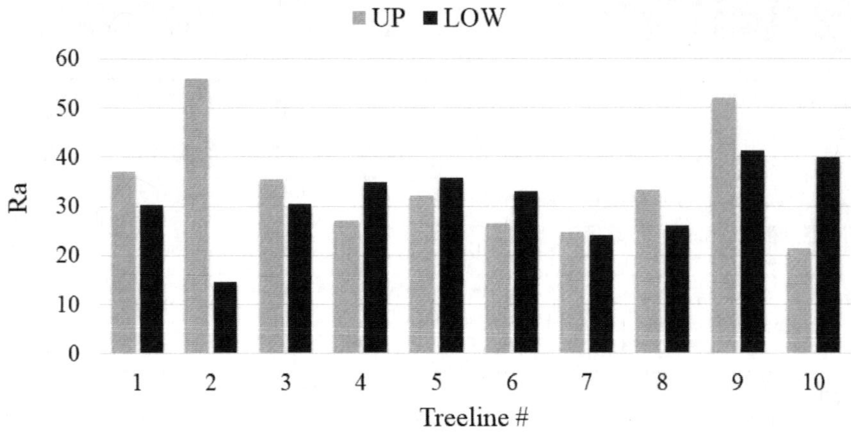

Figure 4. Graphical comparison of roughness parameter (R_a) values for each UP and LOW treeline.

Tables 1 and 2 report the resulting parameters' values for each treeline, and the average values computed over the 10 UP and LOW treelines, respectively.

Figure 4 compares the R_a parameter only, measured over both the 10 UP and LOW treelines. In six out of ten cases the UP datasets show greater values of the roughness parameter, than the corresponding LOW datasets. This observation is in line with the assumption that upper treelines are usually more irregular because less affected by anthropic interventions, than lower treelines that usually exhibit a more regular behavior. Figure 5 provides a graphical comparison of the average values of each surface texture parameter introduced in Section 2, computed over all the UP and LOW treelines. Again, it is possible to notice the prevalence of higher values for the parameters averaged over the available UP treelines. These preliminary results motivate the interest in applying the tools designed for surface texture measurement to the quantitative evaluation of treeline patterns too.

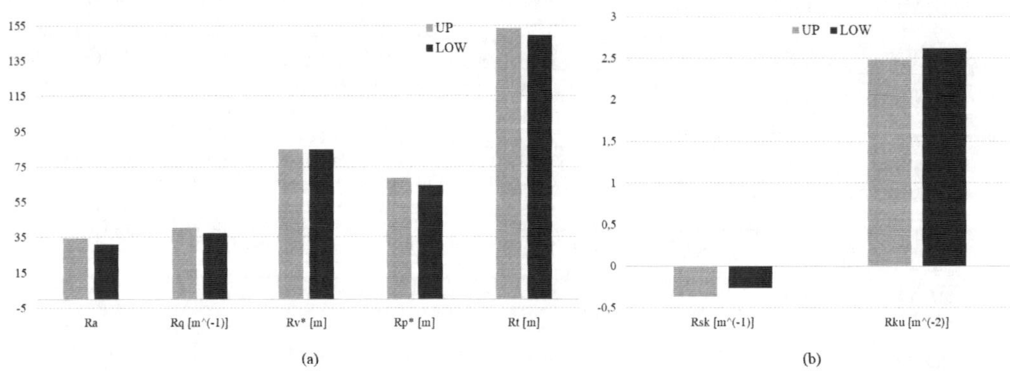

Figure 5. Graphical comparison of average roughness parameters values for all UP and LOW treelines: (a) R_a, R_q, R_v^*, R_p^*, R_t, (b) R_{sk} and R_{ku}.

IOP Conf. Series: Earth and Environmental Science **275** (2019) 012017 doi:10.1088/1755-1315/275/1/012017

4. Conclusion

With the purpose of describing the irregularity of Upper and Lower treelines, this paper proposes a metric based on the analogy to the measurement of surface roughness. Therefore, instead of using interpolated lines, we choose to use the extreme (maxima or minima) points measured; this eliminates all spatial filtering effects caused by interpolation algorithms. Among the available roughness measurements, the R_a one seems more suitable than others to catch and represent the different characteristics of Upper and Lower treelines. However, to reinforce this preliminary conclusion, a more significant evaluation of the R_a parameter requires to extend the analysis to a bigger amount of datasets.

References

[1] C. Körner and J. Paulsen, "A world-wide study of high altitude treeline temperatures," *Journal of Biogeography*, vol. 31, no. 5, pp. 713–732, 2004.

[2] M. A. Harsch, P. E. Hulme, M. S. McGlone, and R. P. Duncan, "Are treelines advancing? A global meta-analysis of treeline response to climate warming," *Ecology Letters*, vol. 12, no. 10, pp. 1040–1049, 2009.

[3] F.-K. Holtmeier and G. Broll, "Treelines - Approaches at Different Scales," *Sustainability*, vol. 9, no. 5, 2017. [Online]. Available: http://www.mdpi.com/2071-1050/9/5/808

[4] R. Daubenmire, "Alpine timberlines in the Americas and their interpretation," *Butler Univ. Bot. Stud.*, vol. 11, no. 8, pp. 119–136, 1954.

[5] ISO/TC 213 Dimensional and geometrical product specifications and verification, "Geometrical product specifications (GPS) - surface texture: Profile method - terms, definitions and surface texture parameters." *ISO 4287:1997*, vol. 04, 1997.

IOP Conf. Series: Earth and Environmental Science **275** (2019) 012018 doi:10.1088/1755-1315/275/1/012018

Scaling-up of evapotranspiration measurements: the promising role of scintillometry

N Vendrame, L Tezza and A Pitacco

Università degli Sudi di Padova - DAFNAE,Viale dell'Università 16, I-35020 Legnaro (PD), Italy

andrea.pitacco@unipd.it

Abstract. The assessment of evapotranspiration (ET) is of primary interest, especially for agricultural and environmental issues. Traditional micrometeorological techniques require homogenous surface and are characterized by limited footprint. Thus, they cannot be applied to assess ET over inhomogeneous and extensive surfaces, typical conditions at the catchment scale and of extensive farming systems. In this context, a suitable technique to measure turbulent fluxes is scintillometry, which can give measurements of sensible heat flux at larger scale, providing averages over heterogeneous surfaces. ET can then be estimated as residual of the energy budget. In this study, we present results from a one-week campaign held during summer 2016 in Southern Italy. We deployed a Large Aperture Scintillometer (LAS) in an extensive vineyard of 140 ha on a path length of 760 m. The site was characterized by gently slope terrain with uniform crop. In order to have reference measurements of local sensible heat flux, we deployed three sonic anemometers along the scintillometer path. The aim of the study was to test the ability of scintillometry to provide a spatially averaged flux, representative of the possibly diverse conditions in an extended footprint upwind to the measurement path. The relationship between sensible heat flux measured by EC and LAS showed to be very good for the EC station in the middle of the path, whereas off-centre areas were less represented.

1. Introduction

Evapotranspiration (*ET*) measurement over inhomogeneous and extensive surfaces – typical conditions at the catchment scale – is still challenging due to spatial variability of vegetation, soil conditions and land topography. Traditional micrometeorological techniques, e.g. eddy covariance (EC), cannot be applied under these conditions, requiring homogenous surface and being characterized by limited footprint. In this context, a suitable technique to measure turbulent fluxes is scintillometry, which can give measurements of sensible heat flux (*H*) at larger scale, providing averages over heterogeneous surfaces. Sensible heat flux is derived from the measurement of turbulent intensity of the air refraction index, in the theoretical framework of Monin-Obukhov similarity theory [1]. Average *ET* over the source area can then be estimated as residual of the surface energy balance [2].

The assessment of *H* and *ET* at large scales is required to validate land surface and meteorological models, but it is also of great interest for extensive farming systems. The first scintillometry measurements over vineyards were performed when the method was still experimental [3]. However, in recent years, with modern and consolidate instruments, application of scintillometry over vineyards is lacking. In this study, we present results from a one-week campaign held during summer 2016 in a vineyard in Southern Italy. The aim of the study was to test the ability of scintillometry to provide a spatially averaged flux, representative of the possibly diverse conditions in an extended footprint upwind to the measurement path.

2. Material and methods

Measurements were carried out from July 31st to August 5th 2016 in the "Tormaresca" estate winery, an extensive private vineyard (*Vitis vinifera*) of 140 ha located in Minervino Murge, South Eastern Italy (41°8′50.08″N, 16°2′33.47″E) (figure 1). The site is characterized by a gentle slope and uniform vegetation, subdivided into different irrigation plots. Sensible heat flux measurements were performed over two irrigation plots of 13 ha (Plot 1) and 6 ha (Plot2) respectively. Both plots are characterized by hedgerow training system with plant spacing of 0.8 m, rows spaced 2.0 m apart and canopy height of 1.8 m. The irrigation was sequentially applied to plots: Plot2 was irrigated from 7:00 for 8 h and Plot1 from 18:00 of the same day for 10 h, with a rate of 1 mm/h.

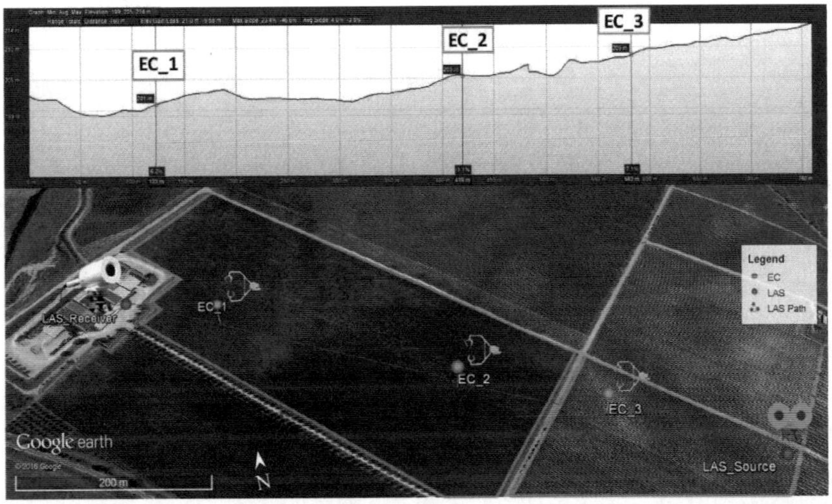

Figure 1. Bottom: satellite image of the study site with Large Aperture Scintillometer (LAS) path (red line); LAS source and receiver positions; EC station positions (*EC_1*, *EC_2*, *EC_3*). Blue contour area: irrigation Plot1; orange contour area: irrigation Plot2. Top: elevation profile along the path and EC station positions.

Area-averaged sensible heat flux over the two plots was determined with a Large Aperture Scintillometer (LAS) BLS900 (Scintec AG, Rottenburg, Germany) on a path length (distance between LAS transmitter and receiver) of 760 m. Raw statistics were calculated every 30 sec and then averaged over 30 min to obtain final fluxes. Data processing was performed using the SRun software vs 1.39 (Scintec AG, Rottenburg, Germany). A crucial input parameter for the calculation of sensible heat flux by LAS is the beam effective height (z_{eff}), which is the distance of the beam from the canopy displacement height (d). If the terrain is undulated, as it is at our site, z_{eff} will vary along the path and an average value should be calculated multiplying the beam distance from d at several points along the path by a specific coefficient obtained from a path weighting function (PWF), as indicated in Scintec Scintillometers Theory Manual, vs 1.04. The PWF describes the relative contribution of turbulent eddies to the average flux in relation to their position on the scintillometer path [4]. We calculated z_{eff} from 10 points along the path, obtaining an average value of 6.4 m.

Local measurements of sensible heat flux were carried out applying the eddy covariance technique at three different points along the LAS path. Two stations (*EC_1* and *EC_2*) were placed in Plot1 and one station (*EC_3*) in Plot2, in order to obtain reference values to compare with the flux by LAS. Each station was equipped with a CSAT3 sonic anemometer (Campbell Scientific, Inc., Logan, UT, USA), where 3D wind components and air temperature were sampled at 20 Hz. In addition, some ancillary meteorological variables (downward/upward shortwave and longwave radiation; soil heat flux, soil temperature and humidity) were measured at *EC_2* station. EC fluxes and other statistics were then calculated on 30–min time intervals.

3. Results and discussion

The timecourse of half-hourly fluxes by EC and LAS during the measurement period is presented in figure 2. The sensible heat fluxes derived from LAS (H_{LAS}) and measured by EC (H_{EC}) showed in general good agreement. Nevertheless, some differences were found among the three EC locations.

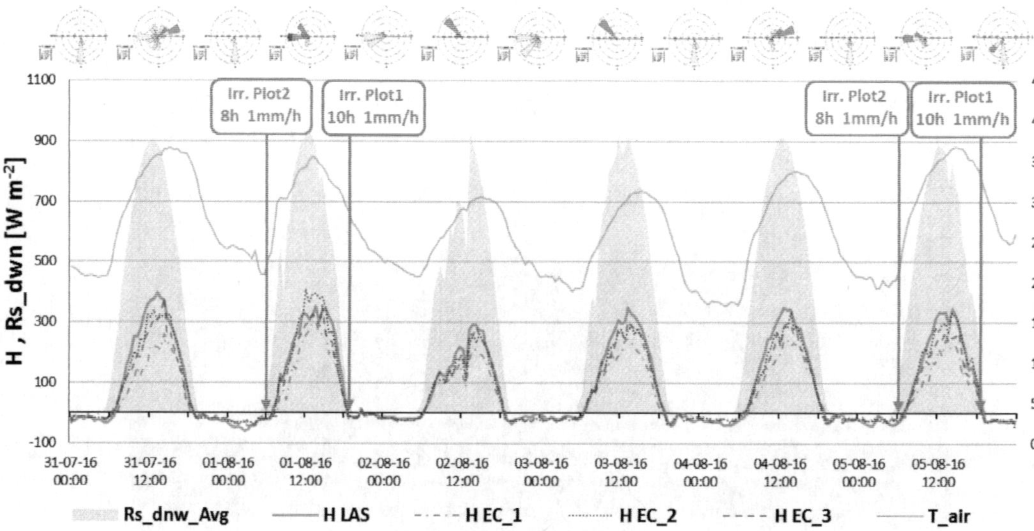

Figure 2. Bottom: timecourse of sensible heat fluxes by LAS (H_{LAS}, light blue thick solid line), EC_1 (H_{EC_1}, green dash dotted line), EC_2 (H_{EC_2}, purple dotted line) and EC_3 (H_{EC_3}, red dashed line); incoming shortwave radiation (yellow area), air temperature (orange thin solid line). Top: nighttime (19:30 – 5:00) and daytime (5:30 – 19:00) wind rose plots during the measurement period.

H_{EC_1} was lower compared to H_{LAS} and the other EC fluxes, even if it was located in the same irrigation plot of *EC_2*. This difference was probably related to the soil characteristics in *EC_1* footprint. In this area the soil is constituted by fine texture with higher water retention capacity compared to the other areas of the vineyard, which would lead to higher evapotranspiration flux and, consequently, lower sensible heat flux. On the contrary, H_{EC_2} and H_{EC_3} were quite similar, except during the days when irrigation was applied. In Plot2 (*EC_3*) irrigation started in the morning, while in Plot1 (*EC_2*) in late afternoon. During the days of irrigation (August 1st and 5th), H_{EC_3} was lower compared to H_{EC_2} as expected, indicating that the evapotranspiration was higher in Plot2.

In general, H_{LAS} was slightly larger than H_{EC} and the best agreement between the two methods was found for the EC station in the middle of the path. This is in accordance with the theoretical PWF for the BLS900, a bell shape curve with a peak in the center of the path [4], meaning that the scintillometer obtains weighted averages instead of arithmetical. However, during the first day of irrigation H_{LAS} was lower than H_{EC_2}, presenting the same pattern of H_{EC_3}, and, on the contrary, during the second day of irrigation H_{LAS} was not reduced, having values very close to H_{EC_2}. This behavior could be explained by the differences in source area for the two days related to wind conditions, even if the prevailing wind direction was the same in both days. During August 1st the wind speed was higher, therefore reducing the footprint area of the scintillometer and increasing the relative importance of Plot2.

4. Conclusions

Large Aperture Scintillometer provided a very reliable estimate of area-averaged sensible heat fluxes over the slightly heterogeneous conditions we have been investigating. However, estimates of sensible heat flux by LAS proved to be very sensitive to the effective beam height used in data processing, therefore a proper knowledge of field morphology is a necessary prerequisite even in gentle slope terrain. Scintillometer proved to be an adequate technique for measuring sensible heat flux at a large scale, which can be used to validate sensible heat by remote sensing methods and improve the evapotranspiration estimation at the farm scale even in heterogeneous conditions.

References

[1] Moene A F, Meijninger W M L, Hartogensis O K, Kohsiek W and de Bruin H A R 2004. A review of the relationships describing the signal of a Large Aperture Scintillometer. Internal Report 2004/2, Wageningen University, Wageningen

[2] Meijninger W M L, Hartogensis O K, Kohsiek W, Hoedjes J C B, Zuurbier R M and De Bruin H A R 2002. Determination of area-averaged sensible heat fluxes with a large aperture scintillometer over a heterogeneous surface – Flevoland field experiment. *Bound.-Lay. Meteorol.* **105** 37–62

[3] De Bruin H A R, Van Den Hurk, B J J M and Kohsiek W 1995. The scintillation method tested over a dry vineyard area. *Bound.-Lay. Meteorol.* **76** 25–40

[4] Wang T, Ochs G R and Clifford S F 1978. A saturation-resistant optical scintillometer to measure C_n^2. *J. Opt. Soc. Am.* **68** 334–338

Acknowledgments

Research funded by the University of Padova, Department of Agronomy, Food, Natural resources, Animals and Environment, grant CDPA148553 (2014). The authors thank the Tormaresca estate winery for hosting the field trial.

Automated crop monitoring solutions to assess the blooming charge in orchards: preliminary results achieved by a prototype mobile lab used on apple trees

G Daglio[1], R Gallo[1], S Petrera[1], C Andergassen[2], M Kelderer[2] and F Mazzetto[1]

[1] Faculty of Science and Technology, Free University of Bozen (FUB), Piazza Universitá 5, 39100 Bolzano (Italy)
[2] VZ-Laimburg, 39040 Post Auer, Südtirol, (Italy)

e-mail: gabriele.daglio@unibz.it

Abstract. The paper presents and discusses the use of a mobile lab (*ByeLab*) equipped with optical sensors (OptRx™ and LiDAR), GNSS- RTK system and an Inertial Measurement Unit (IMU), to carry out a site-specific bloom charge assessment in apple trees. The collected data are firstly synchronized by a LabView software and then processed by dedicated algorithms in MatLab®. Preliminary investigations were performed in order to evaluate the features of the flower and leaves reflectance signature. It was observed that low NDVI values corresponded to the flowers while higher values to the leaves. It was expected that NDVI show a negative correlation with the bloom charge. Using Terrestrial Laser Scanner it was possible to validate the information about the canopy thickness collected by the LiDARs mounted on the ByeLab. The OptRx™ do not seem suitable to be applied to perform an estimation of the flower charge but, merging the data collected by the two types of sensors it is possible to build descriptive maps, which provide information about canopy thickness and vegetation index values. These can be used as reference for automating machines to be applied in thinning operations according to a site-specific approach.

Keywords: Optical sensors, Precision Farming, Ground based sensing, Vegetation Index

1. Introduction

Precision Agricultural (PA) technologies and techniques, such as: remote, proximal and ground sensing, gives many tools to decreasing the production costs and to improve the field yields. Moreover, PA aims to increase the efficiency and the sustainability of agriculture [1].

Ground remote-sensing technologies can be useful to automate several procedures which are generally carried out with techniques that require high amount of time. For example, the use of optical sensors installed on aerial or terrestrial vehicles, was assessed to provide punctual information, about the crop, such as: plants healthy status, grapevine diseases detection, canopy layout and volume and distance between plants [2,3,4].

Flower and fruit thinning are two procedures applied with the aim to improve fruit size and the annual yield apple trees. Usually the flower thinning is carried out through chemical treatments or manually by specialized staff. Chemical thinning does not take into account the variability between individual trees, but is done on a block basis. The hand-thinning is more expensive in terms of money and time spending but may be necessary if chemical thinning proves inadequate [5]. In order to carry out correctly these procedures it is fundamental estimate correctly the blooming charge. Usually this operation is done manually basing upon the personal experience of the operator and it is very complicate to measure the correct number of flowers [6]. Thus, the aim of this work is to develop an automatic system able to provide objective information about the bloom charge in apple orchard to be used as managing tool for flower thinning activities.

2. Material and methods

The experimentation was carried out to the apple orchard of "Azienda sperimentale Laimburg", Bolzano, Italy. Two rows, with plants of *Malus domestica,* were selected. The rows are divided in several transepts trained with alternate system (figure 1): *Bibaum*, having a single trunk with two leaders and *Spindel* with single trunk leader and a circular volume; In both case the plants reach a high of 3.5 m. about.

Figure 1. From left to right: *Bibaum* and *Spindel* training system.

In order to know the real blooming charge of the investigated plants, all flowers were manually counted before starting the experimentation. The data obtained were used to validate the assessment performed by a prototype of mobile laboratory equipped with sensors developed to performed crop monitoring, during the days survey carried out in: pre-blooming, full-blooming and post-blooming in April 2018.

Preliminary investigations were performed before running ByeLab field tests in order to measure the spectral firm which characterized the component of the plant and to be compared with the Vegetation Index (VI) measures got by the optical sensors installed on the ByeLab. To this aim a portable spectrophotometer (Jaz Ocean Optics Spectrometers) was used to assess the different reflectance behavior of different plant organs (petals, corolla, flowers, leaves) and to better evaluate the features of the flower reflectance signature (figure 2). Several vegetation indexes were calculated using RED, RedEdge and NIR wavelengths values of the reflectance in order to define new indexes able to discriminate the flowers from the leaves. In each survey, samples of 30 flowers and leaves were analysed with the spectrophotometer.

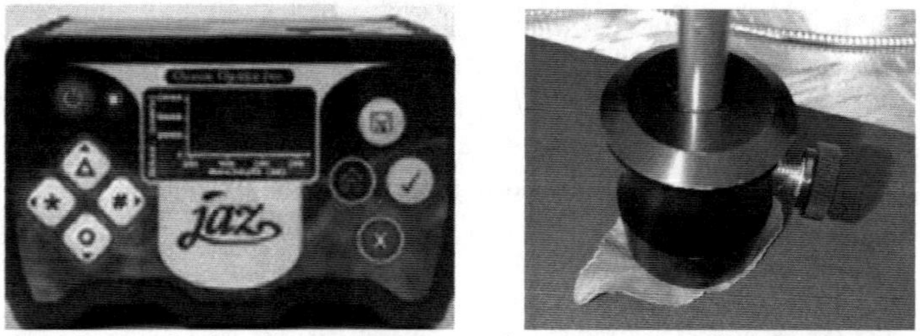

Figure 2. The Jaz Ocean Optics Spectrometers and an example of leaf reflectance acquisition.

Through to the use of a Terrestrial Laser Scanner (TLS) (model CAM2 FARO) it was also possible to validate the information about the canopy thickness collected by the LiDARs mounted on the mobile lab. The collected raw-data were elaborated with FARO Scene 7.1.1 software (figure 3).

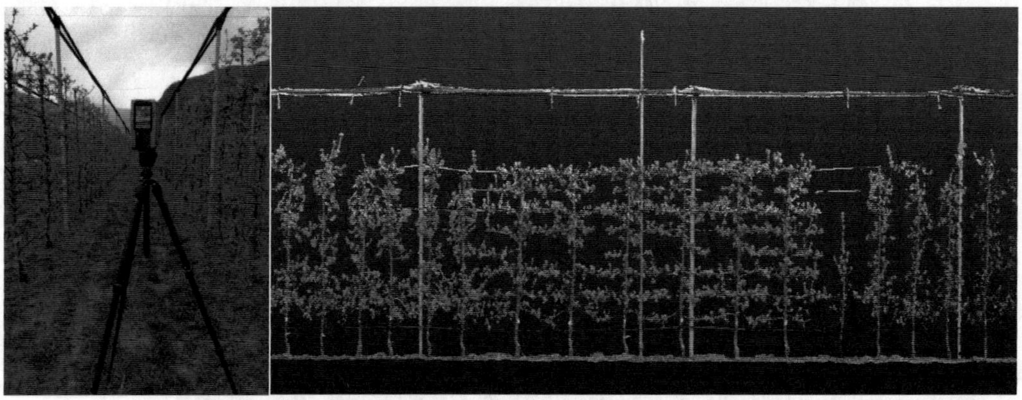

Figure 3. On the left the TLS during the scan and on the right an example of raw data elaborated with FARO Scene 7.1.1 software.

For the monitoring procedure the BionicYELab (ByeLab) was used (figure 4). This vehicle is an compact electric-driven tracked bins-carrier (NEO Alpin by Windegger S.r.l., Lana, Bolzano, Italy) modified with a metal structure where install the sensors and it is remote-controlled. For this experiment it was equipped with the following devices:

- three OptRx™ sensors AgLeadear;
- two LiDAR (SICK LMS111) sensors;
- an Inertial Measurement Unit (IMU) (LMRK 10 AHRS);
- a GNSS system (GEOMAX Zenith 35).

Figure 4. The ByeLab in the apple orchard during a day of surveys.

The OptRx™ devices are active optical sensors able to provide VI value calculated on the reflectance, the two LiDARs are able to detect, in a stereoscopic view, the profile of the canopy while the IMU device is useful to get data on the mobile robot orientation and to correct data collected by the previous sensors because influenced by relevant vibrations generated by terrain roughness. The collected data are firstly synchronized by a LabView® software, running on the control unit, and finally processed by dedicated algorithms in MatLab® environment. In every survey day three repetitions for each semi-canopy were carried out in order to obtain reliable data.

3. Results and discussion

The results obtained by the spectrophotometer highlight that the spectral signature of leaves and flowers (figure 5), even if collected from different plant, have similar patterns but they may have different reflectance values for the same wavelength.

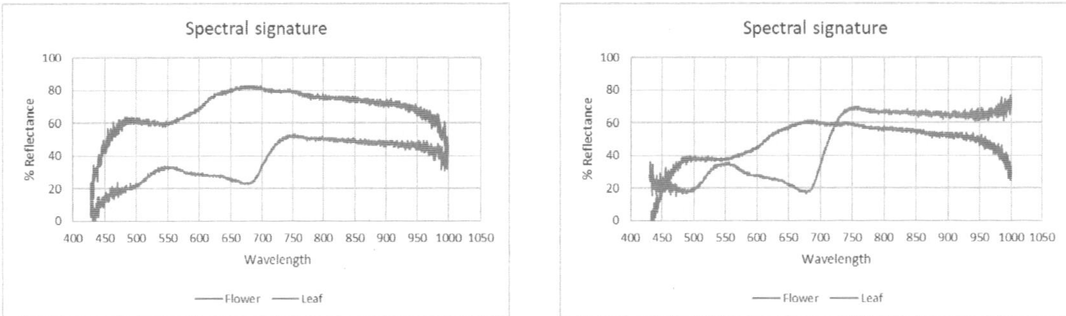

Figure 5. Example of spectral signature of flowers and apple leaves sampled from different trees in full blooming period.

In this work in addition to the most common VIs such as, Normalized Difference Vegetation Index (NDVI) and Normalized Difference Red Edge Index (NDRE), several and new VIs, were tested at the aim to find the most suitable to discriminate leaves than flowers overcoming the issue above cited. Basing to the obtained results it was possible to observe that the NDVI index is the most suitable parameter to be used to discriminate flowers from leaves.

According to the measurements done with the spectrophotometer, it is shown that NDVI index was the best for flower identification. Indeed, the lower the NDVI values, the higher the amount the flowers.

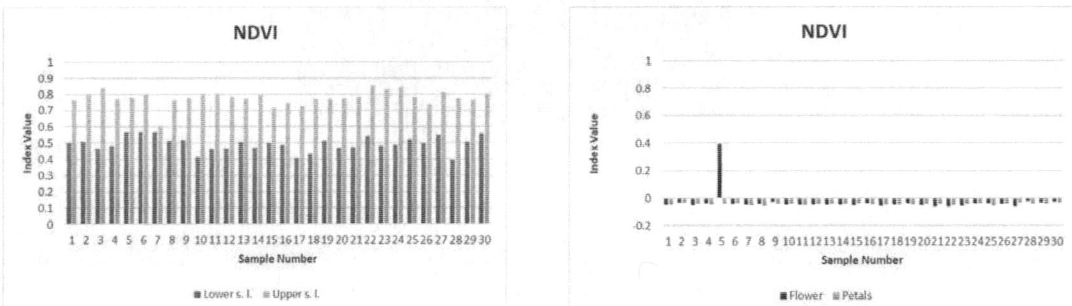

Figure 6. Values of NDVI index collected by the spectrophotometer. In the left picture are shown data related to the leaves (lower and upper surface), while in the right those related to the flowers.

Referring to the obtained results it was possible to observe that the NDVI index is the most suitable parameter to be used to discriminate leaves from flowers. Pure flowers shown NDVI values slightly negative, anyhow very near to zero, while to the leaves corresponds values plenty greater than zero. Basing on these results obtained, by the ByeLab data analysis we expect to obtain a negative correlation between the number of flowers and the NDVI value: low NDVI values in the presence of a high number of flowers and at the contrary high NDVI values with presence of a low flower charge.

Comparing the results obtained by the TLS scans processing with those from of the LiDARs, mounted on the ByeLab during the experimentation (figure 7) it is possible highlight that the system is able to obtain information about: gaps between plants, change in training system, canopy thickness, identification of small and new plants and irrigation poles. However, there is an evident stretching of the images probably due to the fact that there is a time delay in the communication between LiDAR and GPS due to difference in acquisition frequency.

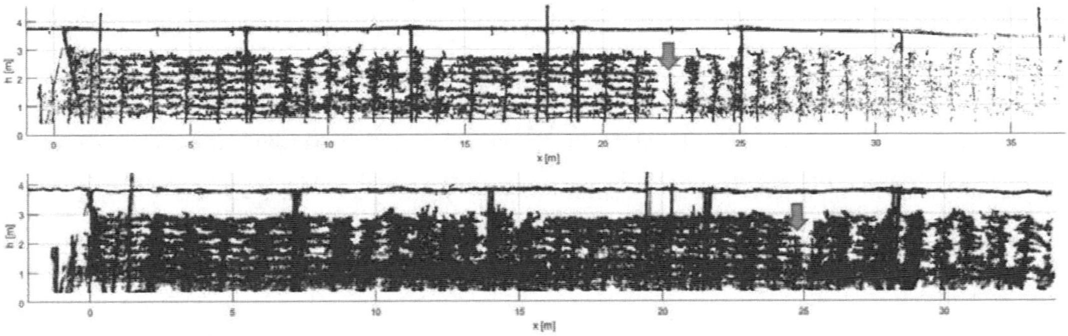

Figure 7. Images obtained by TLS (above) and LiDAR (below). It is observable the stretching between the two acquisitions, the small plant and the information about the canopy.

Thanks to the algorithm elaborations it was possible to obtain descriptive maps (figure 8). These provide information about the blooming charge at three different heights of the canopy. The maps were obtained merging the LiDAR and OptRx® data. Them provide information about and NDVI index values, hence the number of flowers, and the canopy thickness. It was decided to associate cyan and green colors at high NDVI values, while white and yellow colors at low NDVI values. In this way low and high blooming charge were represented, respectively. Since the apple tree is a plant which blooms during the foliage stage the predominant colour in pre-blooming map is cyan. In this stage the leaves are still small and cannot be clearly sensed by the sensors. In full-blooming, since the flower is complete opened there is a predominance of white and yellow, meanwhile it is green in post-blooming because all flowers are felt done.

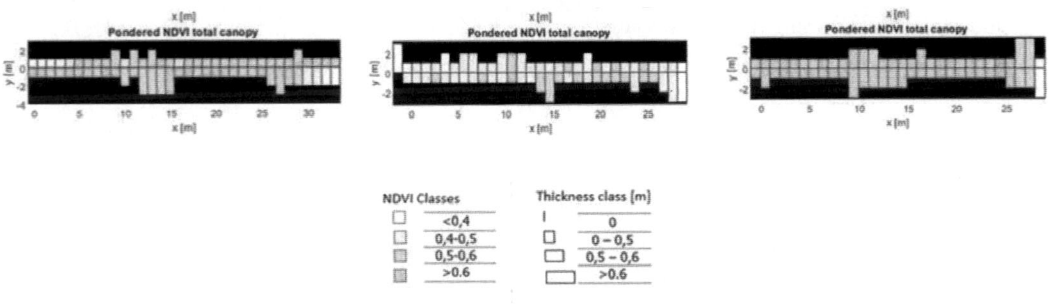

Figure 8. Descriptive maps associated to one scanned row: NDVI and classes thickness classes during the blooming period. From left to right: pre-blooming, full-blooming and post-blooming. The image refers to the total high of the canopy wall.

The descriptive maps shown that the NDVI trend is always low in full blooming (due to the higher amount of white provided by the flowers) if compared with ones in pre and post blooming. However, the variation trend of the NDVI index is really too scarce to set vegetative index thresholds to be used as reference to estimate the density of blooms.

4. Conclusion

According to the results obtained from ByeLab data elaborations, the used of mono-dimensional sensors, such as the OptRx™, does not seem suitable to be applied to perform an estimation of the flower charge. This is due to the low resolution which characterized them: their output refers to an average value of a scanned surface. Two-dimensional scans, for example a pixel matrix, such as multi-cameras, should be used to overcome the problem of low resolution.

Thanks to the combination of LIDAR and OptRx™ data it is possible to provide useful descriptive maps which can be suitable to be used as reference for automatic activation of machines (sprayers) and pruning activities.

In the further studies, Authors will be increased the number of optical sensors per side in order to obtained more reliable results. Beside this, new interpretative algorithms will be developed for the automatic bloom charge assessment.

References

[1] Vidoni R, Gallo R, Ristorto G, Carabin G, Mazzetto F, Scalera L and Gasparetto A 2017 ByeLab: an agricultural mobile robot prototype for proximal sensing and precision farming *ASME 2017 Int. Mech. Eng. Cong.* and Expo American Society of Mechanical Engineers

[2] Rosell J R *et al* 2009, Obtaining the three-dimensional structure of tree orchards from remote 2D terrestrial LIDAR scanning, *Agric. For. Meteorol.,* vol. 149, pp. 1505-15

[3] Gallo R, Ristorto G, Daglio G, Massa N, Berta G, Lazzari M and Mazzetto F 2017 New solutions for the automatic early detection of diseases in vineyards through ground sensing approaches integrating lidar and optical sensors *Chem. Eng. Trans* 58, 673-78

[4] Ristorto G, Gallo R, Gasparetto A, Scalera L, Vidoni R and Mazzetto F 2017 A mobile laboratory for orchard health status monitoring in precision farming *Chem. Eng. Trans* 58, 661-66

[5] Aggelopoulou K D, Wulfsohn D, Fountas S, Gemtos T A, Nanos G. D and Blackmore S 2010 Spatial variation in yield and quality in a small apple orchard *Precis. Agric.* vol. 11, no. 5, pp. 538-56

[6] Xiao C, Zheng L, Sun H, Zhang Y and Li M, 2014 Estimation of the apple flowers based on aerial multispectral image. 2014 ASABE *Annu. Int. Meet. Paper* No. 141912593. St. Joseph, Mich.: ASABE

IOP Conf. Series: Earth and Environmental Science **275** (2019) 012020 doi:10.1088/1755-1315/275/1/012020

Influence of sensor calibration on forecasting models for vineyard disease detection

F Sanna[1,3,*], R Deboli[1], A Calvo[2] and A Merlone[3]

[1] IMAMOTER-CNR - Istituto per le Macchine Agricole e Movimento Terra – Consiglio Nazionale Ricerche, Strada delle Cacce, 73, 10135, Torino, Italy
[2] DiSAFA Dipartimento di Scienze Agrarie, Forestali e Alimentari, Università degli Studi di Torino, Largo Paolo Braccini, 2, 10095 Grugliasco (TO), Italy
[3] INRiM – Istituto Nazionale di Ricerca Metrologica di Torino, Strada delle Cacce 91, 10135, Torino, Italy
Corresponding author: f.sanna@ima.to.cnr.it

Abstract. The evolution of vineyard diseases, such as downy mildew, are depending by temperature, humidity and rain. Fungicides are used to control these pathologies, with considerable economic costs, negative effects on environment, human health and wine quality. In order to identify fungicide sprays periods, several forecasting models were proposed. These tools require accurate knowledge of meteorological variables. These models give great contribution to farmers and technicians, nevertheless, do not consider the quality of the input data in terms of evaluation of measurement uncertainty and traceability to the reference sensors. The inclusion of the sensors' calibration and the influence of weather instrument positioning, affects the disease prediction up to 5 days. Therefore, the choice of instrument position and calibration procedure becomes a matter of importance in agriculture.

1. Introduction

The evolution of vineyard diseases, such as downy mildew, are depending by temperature (T), relative humidity (RH) and rain [1]. The pathologies are controlled with the use of fungicides, which has considerable economic costs, negative effects on environment, human health and wine quality [2,3]. Accurate meteorological data are needed to better predict vineyards diseases. Moreover, Vineyards and other agricultural sites are often positioned on slopes and close to forests where the canopy influences weather conditions in the vicinity. This enforces a non-ideal position for installing weather instruments and the resulting data do not take into account the effect of slope, the proximity of trees or intensity of solar radiation [4]. Generally, the contributions of measurement uncertainties arising from these conditions are not considered. Metrological approach should be applied on agricultural studies.

1.1. Metrology.

As stated by the *Bureau International des Poids et Mesures* (BIPM), the metrology is "*The science of measurement, embracing both experimental and theoretical determinations at any level of uncertainty in any field of science and technology*" [5]. A task of the BIPM as well of the Metrology Institutes is to ensure worldwide uniformity of measurements and their traceability to the International System of Units (SI). Traceability, calibration and uncertainty are fundamental concepts in metrology.

1.1.1. Traceability. Involves the chain of measurements and accuracy transfers that connect the national standards, as maintained by National Metrology Institutes, with the measurements made in research, manufacturing and in field. Traceability is thus defined as the property of the result of a measurement

IOP Conf. Series: Earth and Environmental Science **275** (2019) 012020 doi:10.1088/1755-1315/275/1/012020

related to references through an unbroken chain of comparisons all having stated uncertainties. Traceability makes possible the comparison of accuracy of measurements worldwide, according to a standardized procedure for estimating measurement uncertainty.

1.1.2. Calibration. The term calibration means the act of comparison of an instrument to a reference standard with a known uncertainty and accuracy, and does not include any subsequent adjustment of the instrument. Accuracy is defined as an agreement between the measured value and the true value. The precision instead is the closeness of multiple measurements values. Thus, an instrument can be precise but not accurate.

1.1.3. Uncertainty. The Guide to the expression of Uncertainty in Measurement (GUM), define the uncertainty as the "Parameter, associated with the result of a measurement, that characterizes the dispersion of the values that could reasonably be attributed to the measurand" [6,7].
The GUM publishes by BIPM, asses that the word "uncertainty" means doubt, and thus in its broadest sense "uncertainty of measurement" means doubt about the validity of the result of a measurement. Therefore, we cannot take as valid the value measured by an instrument if this parameter is not associate with it. The parameter can be, for example, a standard deviation called standard uncertainty $u(x)$ of an interval having a probability of coverage established. The coverage factor k is a numerical factor typically in the range 2, where it is assumed that the values fall within an interval having a level of confidence of approximately 95 %. The uncertainty in not an error. Error is the difference between the measured value and the true value of the quantity being measured.

1.1.4. Metrology for agriculture. Since the quality of the measurements is determined by the development of metrology for the specific sector, the metrology can be usefully applied in agriculture field. Indeed, the main objectives of this research are:

- Improving both the meteorological observations and the forecasting models, achieved by inclusion of measurement uncertainties in the input values
- Disseminating techniques, methods and developing procedures for the calibration of weather stations sensors.
- The installation in the agricultural sites of calibrated and traceable automatic weather stations (AWS) and the evaluation of the uncertainty that may bring to optimize the use of pesticides with a positive impact on the environment, health and crops, towards a better management of agricultural risks.

These objectives are linked on the availability of reliable and accurate weather and climate measurements. Weather and climate data are also used in the Decision Support System (such as informational systems that supports farms, companies based on decision-making software). These systems were developed also in compliance with the European Directive n. 128 [8] that established rules for the use of pesticides in order to reduce the risks to human health and the environment.

1.2. Forecasting models

One important rule of the European Directive is the improvement of the forecasting models. These are tools able to transform the relations between crop, disease and the surrounding environment into mathematical equations. In general, the models provide information about the onset and evolution of a disease and alert farmers and technicians to a risk, so treatment can take on time. These tools require accurate knowledge of meteorological variables. Nevertheless, these models do not consider the quality of the input data in terms of evaluation of measurement uncertainty and traceability to the reference sensors [9]. Calibration of weather stations installed in agricultural sites is usually performed by comparison, positioning the reference sensors for a short period close to the station under calibration. This procedure was metrologically evaluated and showed relevant weak points [10,11].

Forecasting models have been developed world-wide in the last decades [12], in particular for diseases affecting viticulture, such as the grapevine downy mildew, an infection strictly depending by temperature, humidity and rain, caused by the fungus *Plasmopara viticola* [1].
The disease is defined by two kinds of infection: the primary infection will occur when specific conditions take place such as temperature at 10 °C or higher, at least 10 mm of precipitation over a 24-hour period. Spores released from the leaf litter germinates inside the leaves forming the characteristic oil spots symptoms. The secondary infection occurs during warm nights where the temperature is over 14 °C, humidity is greater than 90 %rh. The fungus appears as a white mould and spread still on grapes.

1.3. Agrometeorology

The objective of agrometeorology, the branch of meteorology that deals with the relationships among weather and climate on crop, soil management and the environment, as stated in the Guide n. 134 of World Meteorological Organization (WMO) [13], is to assist farmers the agricultural practices through agrometeorological services. The meteorology for agriculture – agro-meteorology – can play a significant role in reducing the negative impacts caused by pests and diseases. An appropriate, preferably integrated, pest management system using meteorological data can reduce losses appreciably.
There is a need for testing various types of sensors, their calibration, and to evaluate the measurement uncertainty related the meteorological quantities [14] in order to improve vineyard disease predictions and reduce the use of chemicals in agriculture.

2. Materials and Methods

Two AWS were installed in a vineyard located in Monferrato (North Italy) and the outcomes data were analysed metrologically in order to evaluate the uncertainty related to the positioning of the sensors and the influence on meteorological measurements. The meteorological data were also used as input values for an epidemiological forecasting model.
The two selected AWS (VA and VB) were specific for agricultural purpose and it composed by the following sensors: air temperature and relative air humidity (combined as thermo-hygrometer), rain gauge, solar radiation, soil temperature and soil moisture, wind speed and direction. The thermo-hygrometer was calibrated in the laboratories of the Italian Metrology Institute (INRiM).

2.1. Sensor's calibration

The calibration results for the temperature and relative humidity sensors are shown in Table 1 and Table 2, and in Figure 1a and 2b. The calibration curves (t_{calc} for temperature RH_{calc} for relative humidity) were obtained by applying a quadratic polynomial function to the data gathered from the sensor in calibration and the reference sensor (t_c), following the equation 1:

$$t_{calc} = \boldsymbol{a}\ t_{AWS}{}^2 + (\boldsymbol{b} + 1)t_{AWS} + \boldsymbol{c} \quad (1)$$

Where: t_{calc} is the calculated data, t_{AWS} is the sensors under calibration and $\boldsymbol{a}$, $\boldsymbol{b}$ and $\boldsymbol{c}$ are the coefficients of the polynomial fitted equation.

The expanded uncertainties (U_t *and* U_{RH}) is expressed as standard uncertainty multiplied by the coverage factor $k = 2$. The U_t *and* U_{RH} evaluated were lower than the calibration target uncertainties proposed for the temperature and humidity, that was in the order of 0.3 °C and 5%, respectively. The statistical contribution $\boldsymbol{U_t}$ was calculated following the Equation 2:

$$\boldsymbol{U_t} = \left(\frac{\sum(t_{calc} - t_c)^2}{\boldsymbol{d}}\right)^{1/2} \quad (2)$$

Where: t_c is the reference sensor, (t_{calc} - t_c) is a residue (the differences between the measured values and those calculated with the polynomial function) and $\boldsymbol{d}$ the degrees of freedom.

Table 1. Components of the calibration uncertainty budget for the temperature sensors

Source of uncertainty budget (T)	Contribute (°C)
Resolution of sensor in calibration	0.1
Uncertainty related to the position	0.03
Reference sensor uncertainty (t_c)	0.01
Interpolation uncertainty	0.035
Standard uncertainty U_t	0.055
Expanded uncertainty U_t (k = 2)	0.11

Table 2. Components of the calibration uncertainty budget for relative humidity sensors

Source of uncertainty budget (RH)	Contribute (%rh)
Resolution of sensor in calibration	0.1
Repeatability of sensor in calibration	0.08
Reference sensor uncertainty (RH_c)	0.01
Interpolation uncertainty	0.70
Standard uncertainty U_{RH}	1.23
Expanded uncertainty U_{RH} (k = 2)	2.5

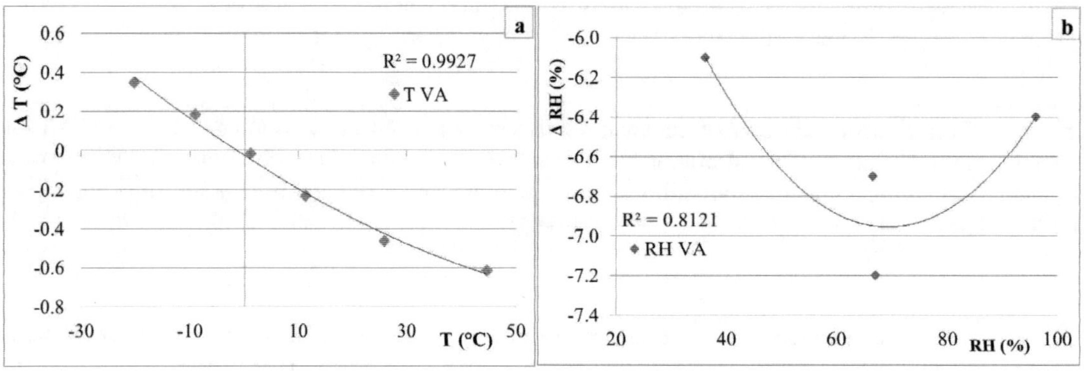

Figure 1. Calibration curve obtained by a polynomial fitted equation for temperature (a) and relative humidity sensors (b) in VA (as example)

2.2. Forecasting model

For this study, the forecasting models choose was the EPI - État Potentiel d'Infection [15], one of the most studied models and still widely used; This model, like others, has improved the quality of the output data, but do not considers the quality of the input data in terms of evaluation of measurement uncertainty and sensors calibration. The model follows the whole life cycle of the pathogen, the EPI index value depends on the sum of a component called "potential energy" that needs climate data and another called "kinetic energy" which use weather data. A risk situation is marked when the index is greater than -10 and this increases constantly in the following three days. The situation risk alerts farmer

on a potential infection, in order to make a treatment in the vineyard. The model equation were validated based on the meteorological trend of the last 15 years, using the data provided by the *Servizio fitosanitario della Regione Piemonte*.
In order to presume the period of spores' germination and the period of primary infection (PPI), the length of incubation period was calculated going backward from the date of primary infection symptoms (DPIS) onset (Figure 2a). An equation temperature and humidity dependent was used (Figure 2b). Taking into account also the precipitations that give the right conditions to spores to spread in the leave, the germination was calculated between the 6 and 7 of April (Table 3).

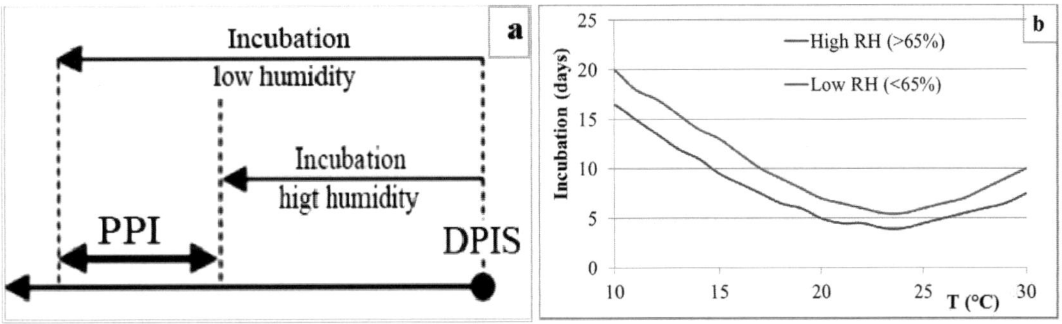

Figure 2. Scheme of period of primary infection (PPI) and date of primary infection symptoms (DPIS) calculation (a); temperature and humidity dependent equation (b).

Table 3. Event that lead to infection (precipitations), period of primary infection (PPI) and date of primary infection symptoms (DPIS) onset.

Precipitation	PPI			DPIS
	Beginning of germination	Sporangia formed	Day of infection	
4 April	6 April	30 April	1 May	9 May
5April	7 April	1 May	1 May	10 May

3. Results

3.1. Simulation and forecast

In Figure 3 and Table 4 are represented the seven scenarios focused on period in which is recommended to make a treatment. The seven simulations were with or without the inclusion, in the input values of the model, of the measurement uncertainties in the upper and lower limits of the confidence interval, for temperature and humidity, and the outcomes are very different among the scenarios. The simulation that better predicted the germination was that one which include uncertainties in upper limit for both temperature and relative humidity.

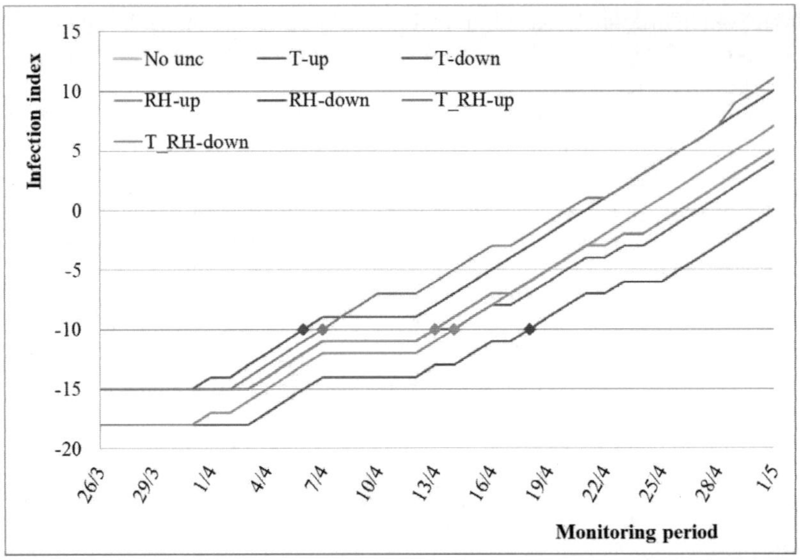

Figure 3. EPI indexes that reached the values -10 for each simulation, the rhombus marks highlighted the day in which is highly recommended starts the treatment.

Table 4. EPI index value focused on period in which is highly recommended make a treatment, data used are without uncertainty (column 2) and with uncertainty in input values (columns 3-4-5-6-7-8). Dark grey boxes indicate the day in which EPI index reached value -10; light grey boxes the days in which the values increases progressively. Red frame cells pointed at the better simulation.

DATA	No unc	T-up	T-down	RH-up	RH-down	T_RH-up	T_RH-down
05-apr	-13	-13	-16	-13	-11	-12	-14
06-apr	-12	-12	-15	-12	-10	-11	-13
07-apr	-11	-11	-14	-11	-9	-10	-12
08-apr	-11	-11	-14	-11	-9	-9	-12
09-apr	-11	-11	-14	-11	-9	-8	-12
10-apr	-11	-11	-14	-11	-9	-7	-12
11-apr	-11	-11	-14	-11	-9	-7	-12
12-apr	-11	-11	-14	-11	-9	-7	-12
13-apr	-10	-10	-13	-10	-8	-6	-11
14-apr	-9	-10	-13	-9	-7	-5	-10
15-apr	-8	-9	-12	-8	-6	-4	-9
16-apr	-7	-8	-11	-7	-5	-3	-8
17-apr	-7	-8	-11	-7	-4	-3	-7
18-apr	-6	-7	-10	-6	-3	-2	-6
19-apr	-5	-6	-9	-5	-2	-1	-5
20-apr	-4	-5	-8	-4	-1	0	-4
21-apr	-3	-4	-7	-3	0	1	-3
22-apr	-3	-4	-7	-3	1	1	-2
23-apr	-2	-3	-6	-2	2	2	-1

If the predictions of the simulation without uncertainties was followed, the pesticide treatments would be carried out a week later, when germination was already begun. Moreover, considering to the potentially washout rains of the following days, the fungicide sprays would have been useless, with significant costs in terms of loss of fungicides, human labour and the final product.

3.2. Position simulation and forecast

Vineyards or other cultures sometimes are positioned on slopes that force a non-ideal positioning of weather instruments. Therefore, we investigated on the temperature and humidity measurements effects on pathogen growth, due to the weather station positioning. In the same vineyard was installed a second calibrated AWS, the first (coded VA) was installed in a sun-exposed place while the second (coded VB) was installed in proximity of trees (approx. 8 and 17 m), where the canopy influenced weather measurements.

Based on the observed symptoms for this second season, the beginning of germination was calculated around 28 and 29 April. The simulations performed were four, considering the two position and data with or without the inclusion of the calibration curve: VA and VB without inclusion of calibration uncertainties in the input values (VA-NC and VB-NC, respectively); VA and VB with inclusion of the calibration uncertainty for temperature and relative humidity (VA-C and VB-C, respectively). A fifth simulation were performed (V-SP) as a control, using the data gathered from an AWS handled by the *Servizio Fitosanitario della Regione Piemonte*, close to the vineyard.

The results of the simulations shown that forecasting without inclusion of the calibration curves predict from four to five days in advance the real spore germination (Figure 5Figure 4), while the simulations with inclusion of calibration curves overlapped at the estimate period of infection.

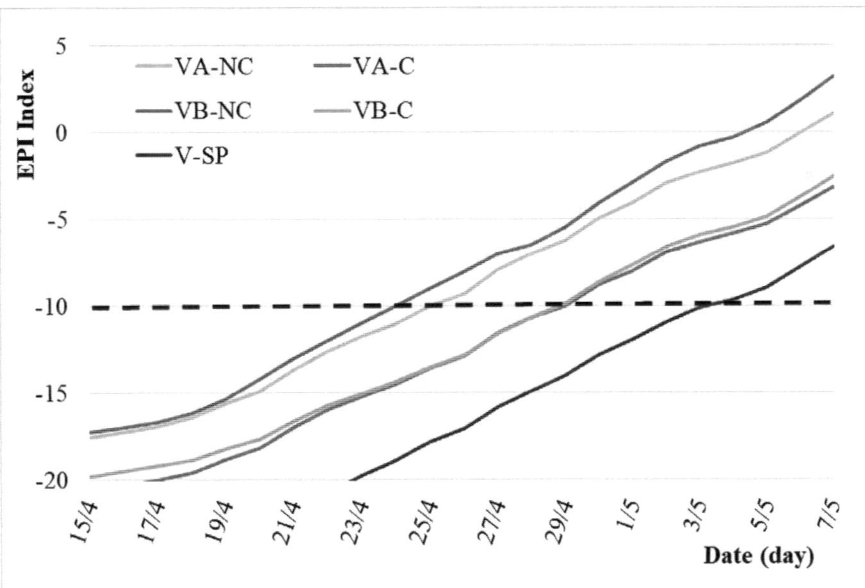

Figure 4. EPI indexes which give estimates of risk for the primary infections. The dashed line is the threshold in which the value reached the critical point -10.

Pesticide treatments carried out when the risk of spores' germination was not occurred yet or was already occurred by a week if we consider the simulation with the data gathered from the phytosanitary service network.

3.3. Secondary infection

In previous studies, the model shown a tendency to overestimate the risk for the secondary infection [16]. In this study, we decide to investigate also this aspect. It could be noticed that using as input values of the model the data from calibrated sensors this overestimation was reduced. Indeed, the symptoms observed around 19-20 June were classified of degree 3 on the base of the European for Plant Protection scale [17], that correspond to a severity of disease of 5-25%.

In the same period, the EPI index produces a value of about 47 and 35 in VB-NC and VA-NC, respectively, compared to the lower values obtained with the calibrated data of 29 and 33 in VA-C and VB-C, respectively (Figure 5). The severities of the disease foreseen are listed in Table 5:

Table 5. EPI index values, percentage of infection and severities of the disease foreseen based on EPPO scale for the four scenarios.

Simulation	Epi index	Infection	EPPO degree
VB-NC	46	67%	5
VA-NC	37	50%	5
VB-C	33	47%	4
VA-C	28	36%	4

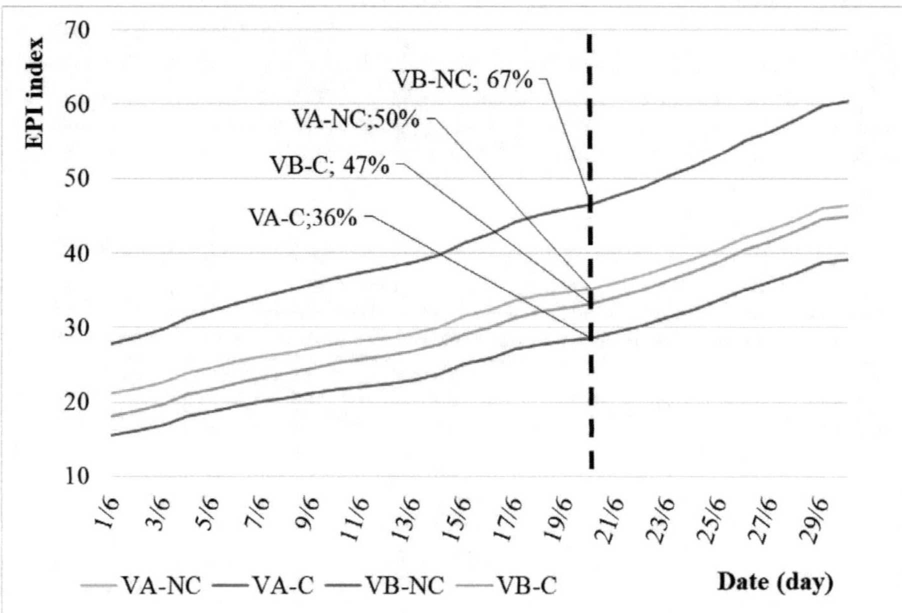

Figure 5. EPI index values and percentage of secondary infection for the four scenarios: VA and VB without inclusion of calibration uncertainties in the input values (VA-NC and VB-NC, respectively); VA and VB with inclusion of the calibration uncertainty (VA-C and VB-C, respectively) for temperature and relative humidity.

The simulation with calibrated data reduced the risk in VA and in VB of 20 and 14%, respectively. Considering the position only, the reduction is at about 11%.

4. Conclusions and Future works

The forecasts provided by calibrated data overlapped the estimated period of infection, confirming that the inclusion of measurement uncertainties produce data closer to the real value of the measurand.
Focusing on the simulations without inclusion of the uncertainties and calibration curves in the input data, the pesticide treatments will be carried out when the diseases was already occurred or not occurred yet, that can be translated in more costs in terms of labour, chemicals waste, with consequences for the human health and environment. Moreover, using accurate data, the overestimation of secondary infection risk was reduced.
The inclusion of the sensors calibration and weather instrument positioning contributions, affects the disease prediction up to 5 days. Therefore, the choice of instrument position and calibration procedure becomes a matter of importance in agriculture. Measurements should be based on fully documented traceability and forecasting models should include measurement uncertainties in the input values, to improve output data reliability

The WMO Guide to Agricultural Meteorological practices n. 134 has valuable information on basic aspects of agrometeorological observations but there is a lack of metrological approach on data analysis, instrument calibration, traceability, and uncertainty evaluations.
The establishment of the metrological requirements for the agrometeorological services and observing systems might allow improving the accuracy in monitoring of local meteorology changes. This is particular true in the vison of a potential creation of a *Reference surface-based observing network*: Measurements performed at different times and sites would be reliably comparable since all measurements would be traceable to the SI.
Metrological approach is needed, also for accurate measurement of soil moisture for a better management of the water supply to the crops as well as for Land-surface temperature and soil temperature. Potential variables to be included in the list of the Essential Climate Variables (ECVs). That critically contributes to the characterization of Earth's climate as defined by the Global Climate Observing System (GCOS).

Acknowledgments
The authors would like to thanks the researcher and technician staff of INRiM and IMAMOTER-CNR for their collaboration and support to this research. This study is funded by the EMRP participating countries within European Association of National Metrology Institutes (EURAMET) and the European Union, in the framework of ENV07 MeteoMet European project.

References

[1] Lafon R and Clerjeau M 1988 Downy mildew *Pearson, R.C., Goheen, A.C. (Eds.), Compendium of Grape Diseases* ed U APS Press , St. Paul, Minnesota p 232
[2] Perazzolli M, Dagostin S, Ferrari A, Elad Y and Pertot I 2008 Induction of systemic resistance against Plasmopara viticola in grapevine by Trichoderma harzianum T39 and benzothiadiazole *Biol. Control* **47** 228–34
[3] Friesland H and Orlandini S 2011 Simulation models of plant pests and diseases *Meteorological Applications for Agriculture - Final report. COST Action* ed P Nejedlik and S Orlandini p 329
[4] Matese A, Crisci A, Di Gennaro S F, Primicerio J, Tomasi D, Marcuzzo P and Guidoni S 2014 Spatial variability of meteorological conditions at different scales in viticulture *Agric. For. Meteorol.* **189–190** 159–67
[5] BIPM 2004 What is metrology? *Celebr. signing Metre Conv. 20 May 1875* **1** 5
[6] BIPM 2008 *Guide to the Expression of Uncertainty of Measurement - Evaluation of measurement data* vol 50
[7] BIPM JCGM 200:2008 2008 The international vocabulary of basic and general terms in metrology *Measurement* **3** 72–6

[8] European Parliament 2009 Directive 2009/128/EC of the European Parliament and the Council of 21 October 2009 establishing a framework for Community action to achieve the sustainable use of pesticides *October* **309** 71–86

[9] Sanna F, Calvo A, Deboli R and Merlone A 2018 Vineyard diseases detection: a case study on the influence of weather instruments' calibration and positioning *Meteorol. Appl.* **25** 228–35

[10] Sanna F, Cossu Q A, Roggero G, Bellagarda S and Merlone A 2014 Evaluation of EPI forecasting model for grapevine infection with inclusion of uncertainty in input value and traceable calibration *Ital. J. Agrometeorol.* **19** 33–44

[11] Sanna F, Deboli R and Calvo A 2018 Variability of tomato in protected environment in response to meteorological parameters *Plant, Soil Environ.* **64** 247–54

[12] Rossi V, Caffi T, Giosuè S and Bugiani R 2008 A mechanistic model simulating primary infections of downy mildew in grapevine *Ecol. Modell.* **212** 480–91

[13] WMO-No. 134 2010 *Guide to Agricultural Meteorological Practices* ed World Meteorological Organization (WMO)

[14] Merlone A, Sanna F, Beges G, Bell S, Beltramino G, Bojkovski J, Brunet M, Del Campo D, Castrillo A, Chiodo N, Colli M, Coppa G, Cuccaro R, Dobre M, Drnovsek J, Ebert V, Fernicola V, Garcia-Benadí A, Garcia-Izquierdo C, Gardiner T, Georgin E, Gonzalez A, Groselj D, Heinonen M, Hernandez S, Högström R, Hudoklin D, Kalemci M, Kowal A, Lanza L, Miao P, Musacchio C, Nielsen J, Nogueras-Cervera M, Oguz Aytekin S, Pavlasek P, De Podesta M, Rasmussen M K, Del-Río-Fernández J, Rosso L, Sairanen H, Salminen J, Sestan D, Šindelářová L, Smorgon D, Sparasci F, Strnad R, Underwood R, Uytun A and Voldan M 2018 The MeteoMet2 project - Highlights and results *Meas. Sci. Technol.* **29**

[15] Stryzik S 1983 *Modèle d'état potentiel d'infection: application a Plasmopara viticola* ed Maison Nationale des Eleveurs (Association de Coordination Technique Agricole) **1** 1-46

[16] Vercesi A, Toffolatti S L, Sordi D, Pedrazzini A, Parisi N and Venturini G 2012 Verso una gestione razionale della difesa antiperonosporica in vigneto *Quad. di Ric.* **145** 24

[17] EPPO - European and Mediterranean Plant Protection Organization 2000 Guidelines for the biological evaluation of pesticides, Efficacy evaluation of fungicides & bactericides 47–9

Setting of a precision farming robotic laboratory for cropping system sustainability and food safety and security: preliminary results

R Orsini[1], D Basili[2], M Belletti[1], D Bentivoglio[1], C A Bozzi[3], S Chiappini[3], C Conti[2], A Galli[1], E Giorgini[2], M Fiorentini[1], E S Malinverni[3], A Mancini[4], L Mazzanti[5], E Monaci[1], G Passerini[6], C Pro[2], R Santilocchi[1], A Vignini[5], S Zenobi[1], P Zingaretti[4]

[1]: *Università Politecnica delle Marche - Dept. of Agricultural, Food and Environmental Sciences*
[2]: *Università Politecnica delle Marche - Dept. of Life and Environmental Sciences*
[3]: *Università Politecnica delle Marche - Dept. of Construction, Civil Engineering and Architecture*
[4]: *Università Politecnica delle Marche - Dept. of Information Engineering*
[5]: *Università Politecnica delle Marche - Dept. of Clinical Sciences, Section of Biochemistry, Biology and Physics*
[6]: *Università Politecnica delle Marche - Dept. of Industrial Engineering and Mathematical Sciences*

Corresponding Author E-mail: r.orsini@univpm.it

Abstract. The acceleration of Digital Agriculture is evident through the increased adoption of digital technologies on farms including smart machines, sensors and cloud computing. In this paper we present the preliminary results of the research project funded by Università Politecnica delle Marche in 2018 "PFRLab: Setting of a precision farming robotic laboratory for cropping system sustainability and food safety and security", which is still underway. In this context, as first result, an interdepartmental Research and Services Center called "Smart Farming" has been set up with the aim to strengthen multidisciplinary collaborations in the fields of Agriculture and Forestry, Geomatics, ICT and Robotics. Regarding field activities the SPAD 502 as well as Normalized Difference Vegetation Index (NDVI) provide a good estimate of the Chlorophylla+b content in durum wheat leaves so can be used to predict in a quickly and non-destructively way, the crop greenness status and to identify any nutritional deficiencies in real time. Future research activities are certainly needed to fully explore the potentialities of conservation agriculture and precision farming, and to drive the transition process from conventional agriculture to modern conservation agriculture and precision farming techniques. In-depth studies are planned on the combined effect of nitrogen fertilization and soil management on the main production variables of durum wheat in order to evaluate whether specific tools for precision agriculture applications can find significant diffusion even in Mediterranean cereal based cropping systems.

1. Introduction

Technological innovations have had a disruptive impact on society and the economy; computers, internet, social media, smartphones, robotics, sensors and cloud-based processes have revolutionized how society organizes itself and how individuals and groups make decisions and behave [1].

The practice of farming and food production is not immune to this digital revolution and is forecast to undergo significant change in the coming years [2].

The acceleration of Digital Agriculture is evident through the increased adoption of digital technologies on farms including smart machines, sensors and cloud computing [3].

The use of these techniques presents advantages when compared with traditional methods based on manual work, however, there are still a number of challenges to be overcome [4]. Smart farming (SF), based on the incorporation of information and communication technologies into machinery, equipment, and sensors in agricultural production systems, allows a large volume of data and information to be generated with progressive insertion of automation into the process. SF relies on data transmission and the concentration of data in remote storage systems to enable the combination and analysis of various farm data for decision making [5].

Demographic trends, including aging populations and continued migration of people from rural to urban areas, have attracted the attention of researchers, because labor issues may become a scarcity factor in agriculture. In addition to these trends, the intensification of climate change will continue to alter growing conditions, such as the temperature, precipitation and soil moisture, in less predictable ways. SF tools can help reduce these impacts, keep them constant or reduce production costs in agricultural activities, and they can assist in minimizing environmental constraints [6].

It is therefore important that even in rural areas strongly linked to tradition, this approach is proposed. In this paper we present the preliminary results of the research project funded by Università Politecnica delle Marche in 2018 "PFRLab: Setting of a precision farming robotic laboratory for cropping system sustainability and food safety and security", which is still underway.

2. Preliminary results:

2.1. Research and Services Center "SMART FARMING"

The Research and Service Center called "SMART FARMING" is based in the Department of Agricultural, Food and Environmental Sciences (D3A) of the (UNIVPM).

The Center was created to respond to the growing importance that the acquisition and automated processing of data obtained through Geomatics, ICT and Robotics tools are assuming in agriculture and forestry, both in research and in the applications development. The wide technical-scientific implications that characterize the emerging sector of Precision Farming induce to operate through an ever greater integration between the numerous and different disciplines involved. The Italian Ministry of agricultural food, forestry and tourism policies (Mipaaft) highlights the breadth and multidisciplinary inherent in the concept of Precision Farming: "business management (agricultural, forestry and zootechnics) based on the observation, measurement and response of the set of quantitative-qualitative variables inter and intra-field that intervene in the productive order. This in order to define, after analysis of site-specific data, a decision support system for the entire company management, with the aim of optimizing returns with a view to an advanced sustainability of the climatic and environmental, economic, productive type and social ".

The Center is conceived, therefore, as a scientific structure that promotes synergies and collaborations between different competences, in order to promote the integration of knowledge and the enhancement of equipment (UAVs, Multispectral cameras, GIS tools, meteorological stations, soil moisture sensors, etc.) and laboratories (for soil, vegetable, water, pests, etc. analysis) with a view to enhancing research, experimentation and teaching in the field of Precision Farming with a strong push also to the involvement of national and international companies operating in the sector. The specific purposes of the Center are synthesized below:

1. Strengthen multidisciplinary collaborations, both in research and in teaching, between professors and researchers working in the fields of Agriculture and Forestry, Geomatics, ICT and Robotics and promote training and connection initiatives between academic world and citizenship, public bodies, associations and companies, in relation to the issues covered by the Center's activities;
2. promote the Center's participation in national and international research projects on the Precision Farming issues;
3. Implementing at the UNIVPM scale, using the expertise of the Center, a specific equipped laboratory dedicated to "Precision Farming" where to carry out research and experimentation activities.

2.2. SPAD reading and leaf chlorophyll content

Field activities were conducted on a long-term experiment (LTE) described by Seddaiu et al., 2016 [7] consisting on a rainfed 2-years rotation with durum wheat (Triticum durum L.) in rotation with maize (Zea mays L.).

The crop rotation is duplicated in two adjacent fields to allow for all crops to be present each year.

Here we present only the results observed on durum wheat in the year 2018. Within each field, starting from 1994, three tillage (main plot: conventional tillage vs minimum tillage vs no tillage) and three nitrogen fertilizer treatments (sub-plot: 0, 90 and 180 kg N ha^{-1} distributed in two rates) were repeated in the same plots every year and arranged according to a split plot experimental design with two replications.

Before and after each nitrogen fertilization, images were acquired through a drone (mod. 3DR SOLO) equipped with multispectral camera (mod. Red-Edge M - MicaSense).

In order to interpret the imagery and calculate vegetation indices on the leaves of durum wheat were made readings using the chlorophyll meter SPAD 502 (Minolta Camera Co., Osaka, Japan) [8] which assesses the relative greenness of the plants in a rapid, in vivo and non-destructive manner. SPAD -502 is based on the indirect measurement of leaf chlorophyll content. It measures the leaf transmittance in red light at 650 nm (at which chlorophyll absorbs) and in near-infrared light at 940 nm (for the correction of leaf thickness). The readings were taken on the central and distal portion of the 30 randomly selected fully expanded leaves of 30 plants of each sub-plot. After each reading the leaves were removed, hermetically sealed in a plastic bag and immediately refrigerated and then transported to the laboratory where they have been analyzed to determine the chlorophyll content using the Arnon et al., 1949 method [9].

Figure 1 shows the positive relationship between SPAD readings and chlorophyll concentration (mg/g) in durum wheat leaves collected in three different soil managements. SPAD 502 Minolta, according whit Debaeke et al., 2006 [10] is a valid tool for tracing, in a non-destructive way, the chlorophyll content present in the foliar tissues and indirectly for test the nutritional status of the crop.

The adoption of conservative soil management techniques (NT or MT) does not lead depressing effects when compared with conventional techniques opening up interesting prospects for the diffusion of such techniques in the considered environment.

IOP Conf. Series: Earth and Environmental Science **275** (2019) 012021 doi:10.1088/1755-1315/275/1/012021

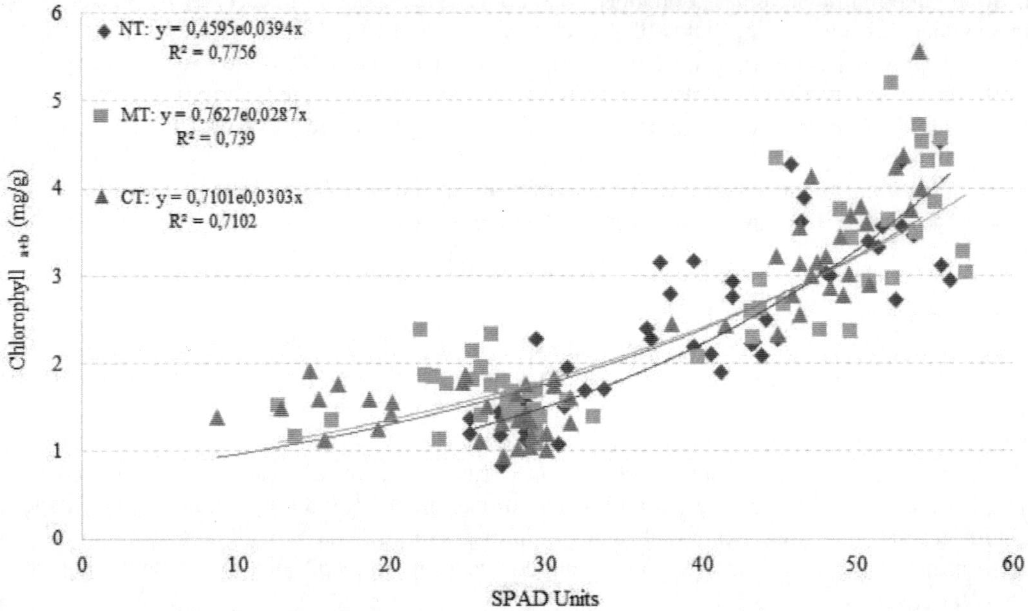

Figure 1: SPAD units and Clorophyll $_{a+b}$ (mg/g) relationship on durum wheat leaves subjected to three different soil managements. Data Points show the average of two replications meausured (NT = no tillage; MT = minimum tillage; CT = Conventional tillage)

2.3 Normalized Difference Vegetation Index (NDVI)

The contemporary long-time satellite remote sensing data provides an advanced way to monitor the surface vegetation dynamics at different spatiotemporal scales [11]. With merits of long continuous time series, good data availability of products based on different remote sensors, the Normalized Difference Vegetation Index (NDVI) is utilized most frequently to analyze the vegetation variations and its correlation with environmental factors in the Northern Hemisphere [12]. NDVI is based on the differential absorption of red and near-infrared spectral bands [13] and quantifies vegetation greenness as related to vegetation vigor and extent [14].

It is strongly correlated to the photosynthetically active radiation absorbed by vegetation, and it is recognized as an indicator of vegetation productivity.

Figure 2 shows the positive relationship between NDVI and chlorophyll concentration (mg/g) in durum wheat leaves subjected to different soil management.

Since NDVI is linked to chlorophyll content, according with Wang et al., 2003 [15], it varies based on vegetation type, the extent of the vegetation cover, the soil, geomorphology and climatic constraints. In the research activities that will be carried out during the 2019, other vegetation indices (i.e. Canopy Chlorophyll Content Index (CCCI), Enhanced Vegetation Index (EVI) [16]) will be calculated and compared to evaluate which is best correlated to the crop nutritional status and to the yield to draw up prescription maps for the contribution with a variable rate of production inputs reducing the potential environmental impact of crop systems.

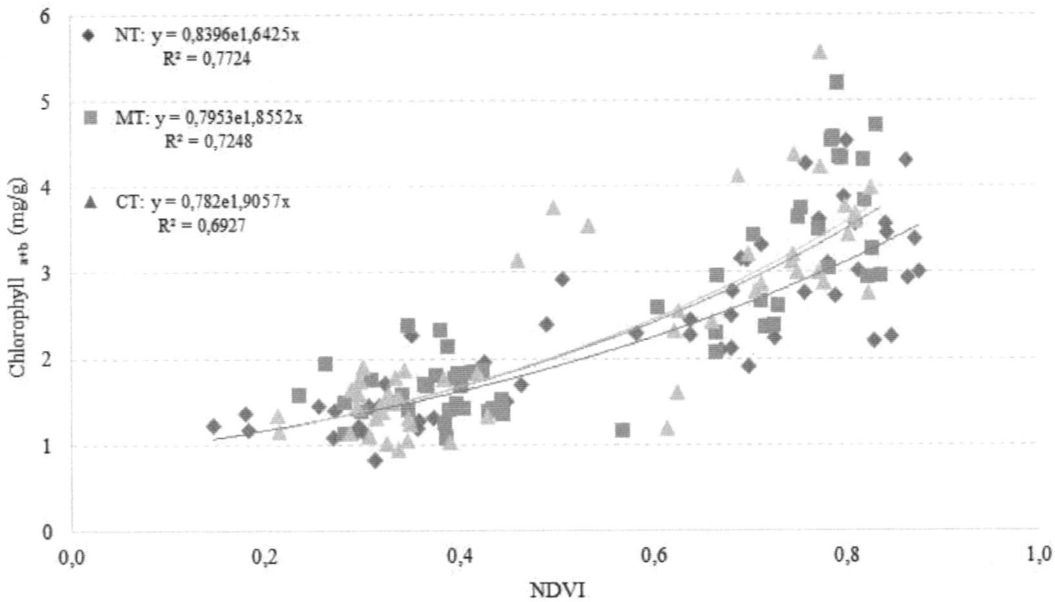

Figure 2: Normalized Difference Vegetation Index (NDVI) and Clorophyll $_{a+b}$ (mg/g) relationship on durum wheat leaves subjected to three different soil managements. Data Points show the average of two replications meausured (NT = no tillage; MT = minimum tillage; CT = Conventional tillage)

3. Conclusions and future perspectives

According to the general objective of Horizon 2020 "to build an economy based on the knowledge and innovation of the European Union, contributing to sustainable development", and following "A strategy for Smart, Sustainable and Inclusive Growth", the project "PFRLab: Setting-up of a Precision Farming Robotic Laboratory for cropping system sustainability and food safety and security" was founded by Università Politecnica delle Marche at the beginning of 2018. As a first result an interdepartmental Research and Services Center called "Smart Farming" has been set up with the aim to strengthen multidisciplinary collaborations in the fields of Agriculture and Forestry, Geomatics, ICT and Robotics and promote training and connection initiatives between academic world and citizenship, public bodies, associations and companies, in relation to the issues covered by the Center's activities.

Regarding field activities emerged encouraging results: the SPAD 502 provides a good estimate of the Chlorophyll$_{a+b}$ content in durum wheat leaves so SPAD 502 can be used to accurately predict in a quickly and non-destructively way, the Chlorophyll$_{a+b}$ for determining the crop nutritional status.

We observed a positive relationship between Normalized Difference Vegetation Index and Chlorophyll$_{a+b}$ content so, NDVI indirectly provides an estimate on the greenness of the crop allowing to identify any nutritional deficiencies in real time using multispectral images acquired by drone.

Given the positive effects that these farming systems can have in the agro-food network (i.e., food security, climate change mitigation, soil fertility, nutrients availability, efficient fertilizer use, low chemicals input), future research is certainly needed to fully explore the potentialities of conservation agriculture and precision farming, and to drive the transition process from conventional agriculture to modern conservation agriculture and precision farming techniques. In-depth studies are planned on the combined effect of nitrogen fertilization and soil management on the main production variables of durum wheat in order to evaluate whether specific tools for precision agriculture applications can find significant diffusion even in Mediterranean cereal based cropping systems.

References

[1] Regan Á 2019 "Smart farming" in Ireland: A risk perception study with key governance actors. NJAS - *Wageningen Journal of Life Sciences* In Press. DOI: https://doi.org/10.1016/j.njas.2019.02.003
[2] Bronson K and Knezevic I 2016 Big data in food and agriculture *Big Data Soc.*, 3 **1**
[3] Carolan M 2016 Publicising Food: big data, precision agriculture, and co-experimental techniques of addition *Sociol. Ruralis*, 57 **2**
[4] Barbedo JGA 2016 A review on the main challenges in automatic plant disease identification based on visible range images *Biosyst. Eng.*, 144 pp 52-60
[5] Pivoto D Dabda P Waquil Talamini E Finocchio CPS Dalla Corte VF Mores GV 2018 Scientific development of smart farming technologies and their application in Brazil *Inf. Processing in Agric.* 1 **5**, pp 21-32
[6] Fountas S Carli G Sørensen CG Tsiropoulos Z Cavalaris C Vatsanidoud A Liakos B Canavari M Wiebensohn J Tisserye B 2015 Farm management information systems: current situation and future perspectives *Comput Electron Agri* **115** pp 40-50
[7] Seddaiu G Iocola I Farina R Orsini R Iezzi G Roggero PP 2016 Long term effects of tillage practices and N fertilization in rainfed Mediterranean cropping systems: durum wheat, sunflower and maize grain yield *Eur. J. Agron.*, **77** pp 166-78
[8] Minolta 1989 Chlorophyll meter SPAD-502. Instruction manual. Minolta Co., Ltd., Radiometric Instruments Operations,Osaka, Japan
[9] Arnon DI 1949 Copper enzymes in isolated chloroplasts. Polyphenoloxidase in Beta vulgaris. *Plant Physiol.* **24** pp 1-15
[10] Debaeke P Rouet P Justes E 2006 Relationship between the normalized SPAD index and the nitrogen nutrition index: Application to durum wheat *J. Plant Nutr.*, **29** pp 75–92
[11] Huete A 2016 Ecology vegetation's responses to climate variability *Nature*, 531 pp 181-2
[12] Lamchin M Lee WK Jeon SW Wang SW Lim CH Song C Sung M 2017 Long-term trend and correlation between vegetation greenness and climate variables in Asia based on satellite data *Sci. Total Environ.*, **618** pp 1089-95
[13] Rouse JHR Schell J Deering D 1973 Monitoring vegetation systems in the great plains with ERTS *Third ERTS Symposium 1973* (NASA SP-351) **1** pp 309-17
[14] Rhee J Im J 2017 Meteorological drought forecasting for ungauged areas based on machine learning: using long-range climate forecast and remote sensing data *Agric. For. Meteorol.*, **237** pp 105-22
[15] Wang J Rich PM Price KP 2003 Temporal responses of NDVI to precipitation and temperature in the central Great Plains, USA *Int. J. Remote Sens.*, 24 **11** pp 2345-64
[16] Cammarano D Fitzgerald GJ Casa R Basso B 2014 Assessing the robustness of vegetation indices to estimate, wheat N in mediterranean environments *Remote Sens.*, **6** pp 2827-2844

Acknowledgments
This study was conducted in the context of the Call for Strategic Projects of the Università Politecnica delle Marche. We are grateful to all those colleagues and students that contributed to the conduction of the field activities. The professionality and technical contribution of the workers of the "Pasquale Rosati" experimental farm of the Università Politecnica delle Marche are also gratefully acknowledged.

IOP Conf. Series: Earth and Environmental Science **275** (2019) 012022 doi:10.1088/1755-1315/275/1/012022

A review on the use of drones for precision agriculture

Pasquale Daponte, Luca De Vito, Luigi Glielmo, Luigi Iannelli, Davide Liuzza, Francesco Picariello and Giuseppe Silano

Department of Engineering, University of Sannio, Piazza Roma, 21, 82100 Benevento, Italy

E-mail: {daponte, devito, glielmo, luigi.iannelli, davide.liuzza, fpicariello, giuseppe.silano}@unisannio.it

Abstract. In recent years, there has been a strong activity in the so-called precision agriculture, particularly the monitoring aspect, not only to improve productivity, but also to meet demand of a growing population. At a large scale, precise monitoring of cultivated fields is a quite challenging task. Therefore, this paper aims to propose a survey on techniques, applied to precision agriculture monitoring, through the use of drones equipped with multispectral, thermal and visible cameras. For each application, the main limitations are highlighted and the parameters to be considered before to perform a flight are reported.

1. Introduction

Farming is facing many economic challenges in terms of productivity and cost-effectiveness, and the increasing labour shortage partly due to the depopulation of rural areas, as well. Among such global challenges, it should be considered the population increase, the urbanization, an increasingly degraded environment, an increasing trend toward consumption of animal proteins changing in food preferences through aging population and migration, and, of course, the climate change [1, 2]. Furthermore, reliable detection, accurate identification and proper quantification of pathogens and other factors affecting both plant and animal health, are critical to be kept under control in order to reduce economic expenditures, trade disruptions and even human health risks. Thus, a more advanced agriculture needs to be set, characterized by the adoption of *ad hoc* production processes, technologies and tools derived from scientific advances, research and development activities.

Precision farming and measurements have already established paradigms in order to increase farm productivity and quality, as well as improving working conditions through reduction of manual labour. All these factors play an important role in making farming sustainable. Also, many modern farmers already use high-tech solutions, e.g., digitally-controlled farm plants and also unmanned aerial vehicles (UAVs) for monitoring and forecasting. Drones are available at affordable prices and are capable of imaging ground data with corresponding geographic locations. That helps the user to have a complete and clearer picture of the ground information. For instance, multispectral and RGB cameras equipped drones offer the advantage of imaging the near infrared portion of the electromagnetic spectrum over the crops, thus providing the crops health conditions [3].

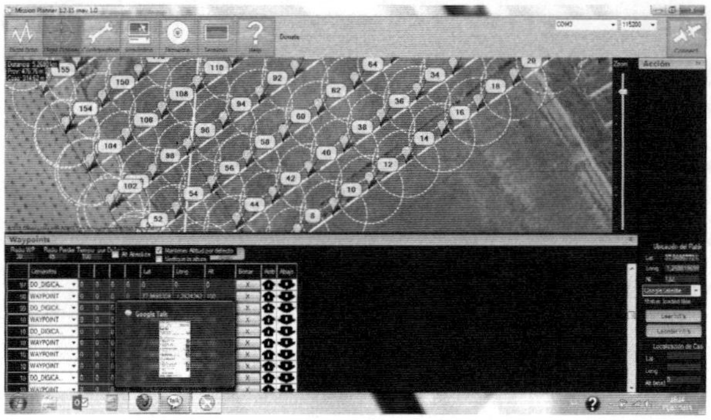

Figure 1. Pre-programmed navigation trajectory for the soil assessment in the APM Planner open-source software.

Drone images and ground sensor data are so expected to play a crucial role in precision agriculture, providing wide room for scientific research and development [4]. Furthermore, several metrological aspects have to be considered for developing such platforms, from the sensors embedded on them up to the instrumentation and the calibration procedures for their testing [5].

Despite their effectiveness and usefulness, the main drawback lies on the fact that these systems are calibrated only for a specific task (e.g., classifying different kinds of vegetation, water bodies, urban, bare soil, etc.), without the ability of creating a holistic view of agricultural processes. This lack of interoperability causes additional work for the human operators, since they have to manually feed the output data from one system to another. For all such reasons, software modules, drones and other equipment are object of research in order to develop a common information middleware and application interface. The aim is to reduce monotonous and time consuming work [6].

In this paper, a review of the drone technology applied to the precision agriculture is presented. In particular, the drone architecture according to the sensors embedded on the payload for precision agriculture applications is reported. Then, the design of a drone in terms of measurements capabilities, power consumption and time of flight according to the application requirements is delineated. Emphasis is also placed in the control part of the drone, understanding its dynamic behavior and how to control it. In such a way, changes in the decision-making system and the mission planner can be made in a simpler way, facilitating the development of different control strategies compared to those already available and validating the effects of modifying the control architecture for complex missions. Finally, the main procedures needed for the calibration of the sensors embedded on drones are reported and an overview of the post-processing tools for extrapolating significant parameters of the monitored area are described.

The rest of the paper is organized as follows. An overview of the drone architectures and the payload sensors for precision agriculture together with its flight control system are reported in Sec. 2 and 3, respectively. Section 4 describes the use of drones for precise agriculture and, for each application, a review of the needed calibration procedures is carried out. Moreover, the main post-processing tools are described. Finally, Section 5 concludes the paper.

2. The architecture of a drone for precision agriculture

As reported in [7], the basic architecture of a drone, without considering the payload sensors, consists of: (i) frame, (ii) brush-less motors, (iii) Electronic Speed Control (ESC) modules, (iv) a control board, (v) an Inertial Navigation System (INS), and (vi) transmitter and receiver

IOP Conf. Series: Earth and Environmental Science **275** (2019) 012022 doi:10.1088/1755-1315/275/1/012022

Figure 2. Architectural overview of the Parrot Bluegrass.

modules.

In precision agriculture, as well as in disaster relief, building inspection or traffic monitoring, the employed drones are semi-autonomous. In that case, the drone has to fly according to the definition of a flight path in terms of waypoints and flight altitude. Thus, the drone has to embed on board a positioning measurement system (e.g., Global Navigation Satellite System, GNSS) for knowing its position with respect to the waypoints. Furthermore, it embeds an altimeter (e.g., barometer, laser altimeter, ultrasonic sensor) for flying at constant flight altitudes. An example of software for defining the mission trajectory is the APM Planner [8]. In Fig. 1, the user interface of this tool is depicted.

The payload of a drone includes all the sensors and actuators that are not used for the control of its flight (e.g., the gimbal with the RGB camera). In case of precision agriculture, the sensors embedded on drones are: multispectral camera, thermal camera, RGB camera and Light Detection and Ranging (LiDAR) systems.

Multispectral cameras are used for quantifying the state of the monitored vegetation in terms of: (i) chlorophyll content, (ii) leaf water content, (iii) ground cover and Leaf Area Index (LAI), and (iv) the Normalized Difference Vegetation Index (NDVI). Thermal cameras have demonstrated high potential for the detection of water stress in crops due to the increased temperature of the stressed vegetation.

For example, in [9], the authors propose a drone for vegetation monitoring using thermal and multispectral cameras. The thermal camera is the thermovision A40M, with 320×240 pixels and having a spectral response in the range of 7.5 µm to 13 µm. Furthermore, each pixel has a resolution of 16 bits and a dynamic range of 233 K to 393 K. The multispectral sensor is the six-band multispectral camera MCA-6 Tetracam. The camera consists of six independent image sensors and optics with user configurable filters, having center wavelengths at 490, 550, 670, 700, 750, and 800 nm, respectively.

RGB cameras and LiDAR systems are usually adopted to digitize the terrain surface to provide the Digital Terrain Model (DTM) or the Digital Surface Model (DSM) of the monitored area. The DSM represents the earth's surface and includes all the objects on it. On the other hand, the DTM represents the ground level of the soil without considering the vegetation height. For example in [10], the authors use the Swinglet CAM drone with embedded the compact camera Canon IXUS 220 HS for estimating the characteristics of a vineyard. They have performed a flight by acquiring the images related to a vineyard in several waypoints. By using the commercial tool Pix4Dmapper [11], the DSM and DTM are extrapolated. In particular, the differential model of the vine rows is obtained by subtracting the DTM from the DSM.

According to the previous examples, in order to use a drone for precision agriculture, at least the following capabilities are needed: (i) the drone has to fly according to waypoints definition,

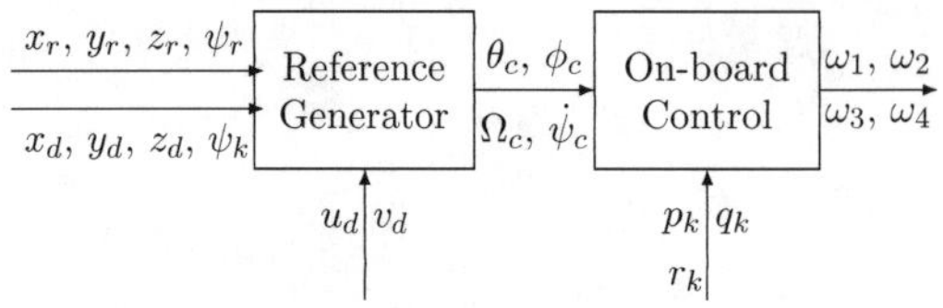

Figure 3. The control scheme. Subscript c indicates the commands, r indicates the references, d indicates the drone variables and k indicates the sensors and data fusion outputs.

(ii) the drone has to control its flight altitude, (iii) the drone has to sense and avoid obstacles during the flight, (iv) the drone has to land according to the state of the battery, automatically and (v) the acquired images have to be stabilized with a gimbal.

An example of drone that can be used for precision agriculture and fulfills the above mentioned requirements is the Parrot Bluegrass (see, Fig. 2). In particular, it can be driven according to preliminary defined waypoints and flight altitude values. Furthermore, it embeds a RGB camera, the Parrot Sequoia multispectral sensor and a sensor for measuring the environmental luminosity.

3. Flight control system

As a common rule in the literature [12, 13], the flight control system of semi-autonomous drones is split into two parts: a reference generator (*outer loop*), that takes into account the position to reach, expressed in terms of waypoint (x_r, y_r and ψ_r) and flight altitude (z_r), and generates the command signals ($\dot{\psi}_c$, Ω_c, θ_c and ϕ_c); and an on-board control system (*inner loop*), that uses such commands providing as output the motors speed ω_i (i from 1 to the number of propellers). They work together in a cascade control structure, where the inner loop (on-board) needs to regulate at a rate faster than the outer loop (that usually runs on a ground control station). Figures 3 and 5 describe the overall system and the on-board control architecture, respectively. The reference generator uses the drone position (x_d, y_d and z_d) and the orientation along z-axis (ψ_k) to compute the command signals ($\dot{\psi}_c$, Ω_c, θ_c and ϕ_c), where the drone position and velocity (u_d, v_d and w_d) come from the positioning measurement system. There are many approaches to carry out the references, from basic heuristic techniques to more advanced model based methods that exploit an accurate dynamical model of the plant. However also classical (and simple) Proportional Integral Derivative (PID) controllers might benefit from a detailed model. In any case, the dynamical model of the aircraft can be easily derived by introducing the so-called inertial (O_{FI}) and body (O_{ABC}) reference systems, as depicted in Fig. 4. Further details can be found in [12, 13].

The aim of the reference generator is to reach the position coordinates (x_r, y_r, z_r and ψ_r) by tuning the desired attitude (θ_c and ϕ_c), the heading velocity ($\dot{\psi}_c$) and the thrust (Ω_c) of the drone, later used as references for the on-board control system. Usually, it is not much complicated to design or modify the available control architecture. For example, in the case of the Parrot Bluegrass, the Software Development Kit (SDK) can be employed to implement a suitable position controller improving the performance of the existing one in the mission planner. Also, the on-board control is decomposed into two parts: the attitude and the rate controller, both illustrated in Fig. 5. The scheme integrates the state estimator that, starting from the accelerometer and gyroscope data, allows to estimate the attitude (ψ_k, θ_k and ϕ_k) and the angular velocities (p_k, q_k and r_k) of the drone, used by the on-board control loop. Usually, such control architecture is "closed", especially when using commercial applications. Therefore, it is not possible to make changes to either the control gains or the control loop (state estimator included). Thus, those drawbacks and limitations have to be considered when going to design a flight control system in a precision agriculture scenario. Disturbances rejection in adverse

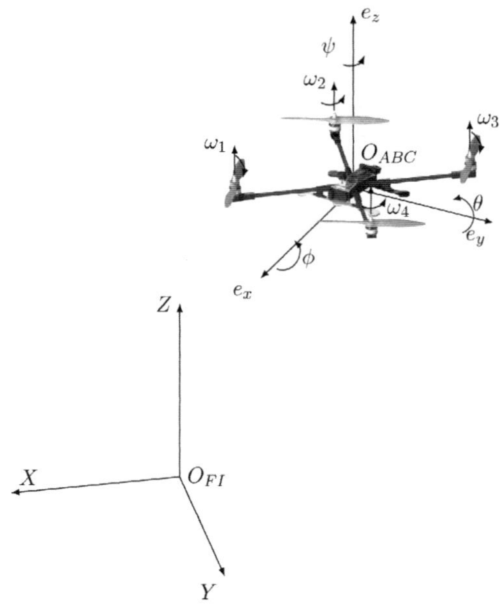

Figure 4. Drone in the body-frame (O_{ABC}) and the fixed-frame (O_{FI}) reference system. The forces, the spin direction and the propellers velocity, ω_i, from each rotor are also reported.

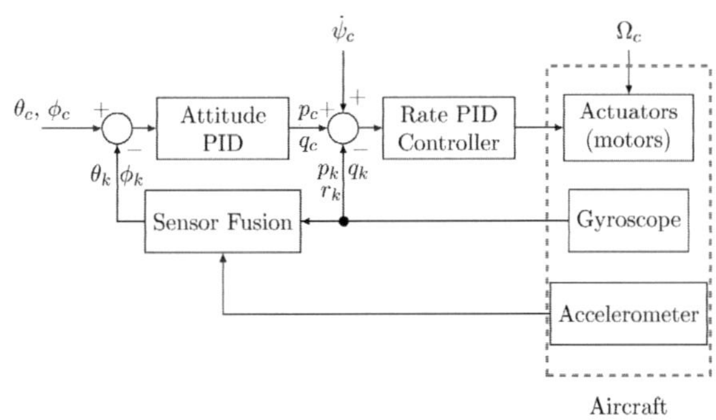

Figure 5. On-board control architecture of a semi-autonomous drone. Usually, the attitude controller runs at 250 Hz while the rate controller runs at 500 Hz.

weather conditions and robustness against model uncertainties are a key feature. Indeed, the aircraft has to be able to control its position under the influence of wind gusts. This is especially true when flying close to obstacles (e.g., threes or vineyards), since position errors due to a wind gust might cause a collision damaging the crops. Current position control methods, such as PID, do not perform well under the influence of gusts. Indeed, PID gust rejection properties scale with magnitudes of the gains, which is often limited by the positioning measurement system update frequency in outdoor scenarios. Moreover, the integrator term is generally slow in compensating persistent wind disturbances. Nowadays, there is much activity of the scientific research in this specific field [14, 15].

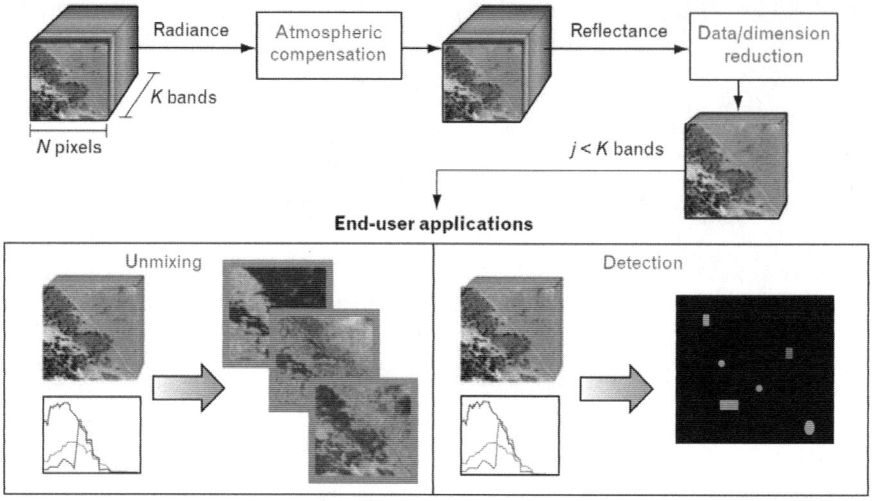

Figure 6. The post processing procedure for extrapolating information related to the state of the vegetation from the acquired multispectral images.

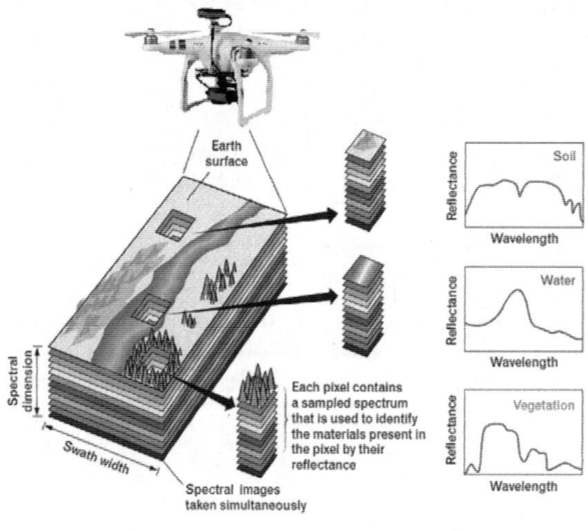

Figure 7. The monitoring operations of a crop by means of multispectral camera embedded on drone.

4. The use of drone for precision agriculture

Intensive agriculture has several negative impacts on the environment. It adds significant and environmentally detrimental amounts of nitrogen and phosphorus to terrestrial ecosystems [16]. Also, excessive fertilizers application can cause pollution risks for the environment, whereas insufficient fertilizer used to replace nitrogen and phosphorus lost through intensive cropping can lead to soil degradation and loss of fertility. Additionally, pollution of water courses and bodies, and consequent degradation of water-related ecosystems are rising due to agricultural chemicals seeping into nearby water. Furthermore, serious soil degradation, which threatens the productivity of the different soils, can be observed all over Europe [17].

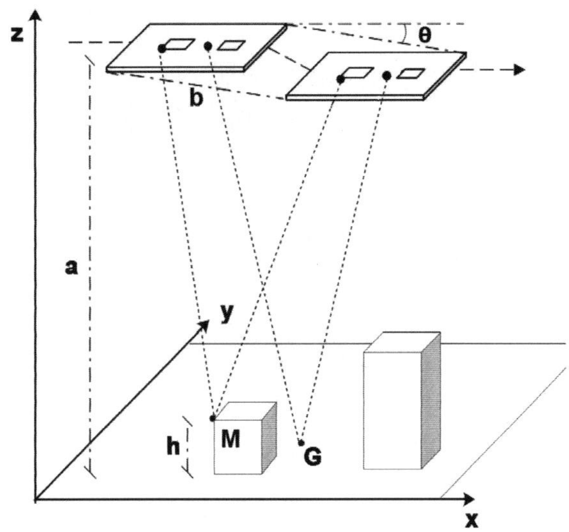

Figure 8. The 3D reconstruction process by considering two images acquired by drone in two waypoints [22].

In addition to the environmental impact, the health risk aspect of the use of chemicals in agriculture needs to be considered. Indeed, chemicals may threat farm workers, as well as families and possibly the inhabitants of the areas surrounding crops, vineyards and farming sites. Additionally, pesticides are absorbed by crop and natural resources (i.e., water and soil) and end up as concealed substances in the food chain, with the increasing risk for both livestock and humans, with huge negative impacts on the public health. Through autonomous precision farming, these effects can be mitigated since chemicals, such as fertilizer and pesticides, are only administered where needed instead of being applied over a large area.

In such a context, the use of drones in agriculture has recently been introduced for big areas inspection and smart targeted irrigation and fertilization [18, 19]. The possibility of detecting, by a drone and an infrared camera, the areas where a major irrigation is needed or where a foliage disease is spreading, can help agronomists to save time, water resources and reduce agrochemical products. At the same time, such advanced farming techniques may lead to increased crop productivity and quality.

Specifically, water deficiency, nutrient stress or diseases can be localized and measured and decision can be made to fix the problem. Many vegetation indexes have been developed which involve various data features, such as the NDVI. Special camera systems are able to acquire data from an invisible part of the electromagnetic spectrum called Near-Infraed (NIR) and extract quality information, such as the presence of algae in the rivers or oil spills near costs [20].

Current agriculture drones applications [21] are: (i) biomasses, crop growth and food quality monitoring, (ii) precision farming, such as to determine the degree of weeds for site-specific herbicide applications, (iii) harvesting and logistic optimization. All these applications require the processing of the images acquired from a camera embedded on the drone.

According to the sensors embedded on drones, it is possible to define three types of applications for precision agriculture: (i) applications based on multispectral and thermal cameras, and (ii) applications based on RGB cameras.

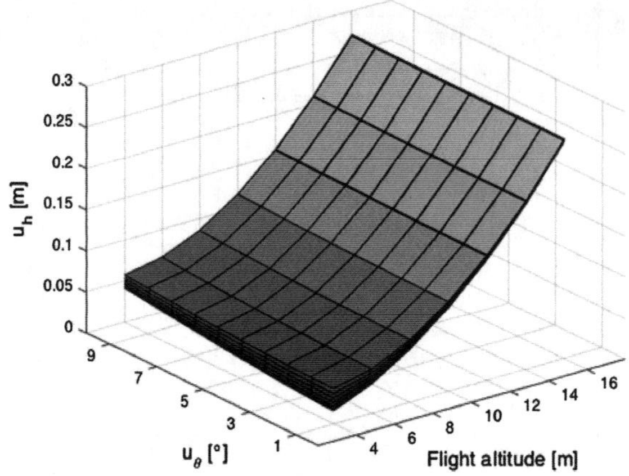

Figure 9. Height uncertainty value u_h vs flight altitude and pitch angle uncertainty value u_θ, for different uncertainty values related to the measurement of the distance between the two waypoints [22].

4.1. Applications based on multispectral and thermal cameras

Usually, for agriculture the terrain is scanned by using satellites with multispectral and thermal cameras. For precision agriculture, due to the needed high spatial resolution, drones are more suitable platforms than satellites for scanning. They offer much greater flexibility in mission planning than satellites.

The drone multispectral and thermal sensors simultaneously sample spectral wavebands over a large area in a ground-based scene (see, Fig. 7). After post-processing, each pixel in the resulting image contains a sampled spectral measurement of the reflectance, which can be interpreted to identify the material present in the scene. In precision agriculture, from the reflectance measurements, it is possible to quantify the chlorophyll absorption, pesticides absorption, water deficiency, nutrient stress or diseases.

There are four sampling operations involved in the collection of spectral image data: spatial, spectral, radiometric, and temporal. The spatial sampling corresponds to the Ground Sample Distance (GSD). The GSD is the distance in meters between two consecutive pixel centers measured on the ground. It depends on the sensor aperture and the flight altitude. The spectral sampling is performed by decomposing the radiance received in each spatial pixel into a finite number of wavebands. The radiometric resolution corresponds to the resolution of the Analog to Digital Converter (ADC) used for sampling the radiance measured in each spectral channel. Furthermore, the temporal sampling refers to the process of collecting multiple spectral images of the same scene in different instants.

Those four sampling operations have to be taken into account for the design of a flight mission and for choosing correctly the multispectral camera and the drone platform. The post-processing procedure required for extrapolating information from the acquired multispectral and thermal images is depicted in Fig. 6. In particular, the images acquired by drone provides measurements related to the radiance in each pixel. In order to measure the reflectance, image processing algorithm are applied to compensate the effects due to the atmosphere absorption and the spectrum of the solar illumination. From the reflectance values, it is possible to detect several materials and the state of the vegetation according to known spectral responses.

4.2. Applications based on RGB cameras

In precision agriculture, the images acquired by drones embedding RGB cameras are used for extrapolating DTM and DSM related to the surveyed area. To this aim, it is important to define the flight mission parameters according to the spatial resolution, and the measurement accuracy of the reconstructed DTM and DSM. As in case of multispectral and thermal cameras, the spatial resolution is defined in terms of GSD. According to the GSD that would be reached, the camera resolution and the flight altitude are chosen. The height measurements of the terrain and of the objects in the scene are obtained by taking two consecutive images from the camera in two different waypoints [22]. The two images have to overlap the same objects in scene (see, Fig. 8). Usually, an overlapping factor of the 70 % between the two images is adopted. From the two acquired images, by knowing the camera parameters, the position and the altitude of the waypoints, it is possible to extrapolate the heights of the objects in the scene.

In [22], the authors propose an uncertainty model for quantifying the accuracy related to the 3D reconstruction of a terrain or a surface by means of aerial photogrammetry. As reported in Fig. 9, they have modeled the uncertainty related to the height measurements for several flight altitudes according to the uncertainty on the measurements of the distance between two waypoints and the orientation of the second waypoint respect to the first one.

From that figure, it is possible to observe that the uncertainty on the height measurements at a flight altitude of 16 m is in the order of 30 cm. Furthermore, they have evaluated the uncertainty related to the commercial tool, Pix4Dmapper [11], adopted to provide DTM and DSM. In that case, at the flight altitude of 11 m, the measurement uncertainty is 30 cm.

From those analyses, it is highlighted that the accuracy related to a 3D reconstruction mainly depends on the accuracy of the drone position measurement system. Thus, according to the target accuracy, several techniques for localizing the drone during the flight can be adopted: (i) differential GPS systems, having a position accuracy in the order of 1 m, (ii) real-time kinematic (RTK) GPS, having an accuracy in the order of 2 cm, and (iii) simultaneous localization and mapping (SLAM) based techniques achieving a position accuracy in the order of 10 cm.

5. Conclusions

In this paper, a review on the use of drones for precision agriculture has been presented. In particular, the general architecture of a drone for multispectral/thermal sensing and DTM/DSM has been discussed. Some technical details about the control system architecture have been also provided. Furthermore, for both the applications, the main limitations and the parameters to take into account before performing a flight are described.

Future trends in this research field go toward the use of cheap commercial mini or micro drones. However, in doing so, the measurement accuracy specifications are challenging to be addressed and several problems arise. For example, for that drones the wind influence, the low GPS accuracy and the strong drift of the INS play destructing effects in flight stability and image acquisition [23].

6. Acknowledgements

This work was partially funded by European Commission H2020 Framework Program under Grant No. 783221 - AFarCloud project - "Aggregate Farming in the Cloud".

References

[1] Gnip P, Charvat K and Krocan M 2008 Analysis of external drivers for agriculture *World conference on agricultural information and IT, LAAID AFITA WCCA* 797–801

[2] Carpenter S R, Caraco N F, Correll D L, Howarth R W, Sharpley A N and Smith V H 1998 Nonpoint pollution of surface waters with phosphorus and nitrogen *Ecological Applications* **8** 559–568

[3] Reinecke M and Prinsloo T 2017 The influence of drone monitoring on crop health and harvest size *2017 1st International Conference on Next Generation Computing Applications* pp 5–10

[4] Murugan D, Garg A, Ahmed T and Singh D 2016 Fusion of drone and satellite data for precision agriculture monitoring *2016 11th International Conference on Industrial and Information Systems* pp 910–914
[5] Daponte P and Debei S 2018 Special issue on selected methods and instrumentation of metrology for aerospace *IEEE Aerospace and Electronic Systems magazine* **33**
[6] Fontanella R, Vetrella A R, Fasano G, Accardo D, Moriello R S L, Angrisani L and Girard R 2017 A standardized approach to derive system specifications for drones operating in the future UTM scenario *2017 5th IEEE International Conference on Models and Technologies for Intelligent Transportation Systems* pp 250–255
[7] Brzozowski B, Daponte P, De Vito L, Lamonaca F, Picariello F, Pompetti M, Tudosa I and Wojtowicz K 2018 A remote-controlled platform for UAS testing *IEEE Aerospace and Electronic Systems Magazine* **33** 48–56
[8] APM planner official web site `http://ardupilot.org/planner/` accessed: Sept. 2018
[9] Berni J A J, Zarco-Tejada P J, Suarez L and Fereres E 2009 Thermal and narrowband multispectral remote sensing for vegetation monitoring from an Unmanned Aerial Vehicle *IEEE Transaction On Geoscience and Remote Sensing* **47** 722–738
[10] Burgos S, Mota M, Noll D and Cannelle B 2015 Use of very high-resolution airborne images to analyse 3D canopy architecture of a vineyard *The International Archives of the Photogrammetry, Remote Sensing and Spatial Information Sciences* pp 399–403
[11] Pix4Dmapper official web site `https://pix4d.com/` accessed: Sept. 2018
[12] Bouabdallah S, Murrieri P and Siegwart R 2005 Towards autonomous indoor micro VTOL *Autonomous Robots* **18** 171–183
[13] Silano G, Aucone E and Iannelli L 2018 CrazyS: A software-in-the-loop platform for the Crazyflie 2.0 nano-quadcopter *2018 26th Mediterranean Conference on Control and Automation* pp 352–357
[14] Smeur E J J, de Croon G and Chu Q 2018 Cascaded incremental nonlinear dynamic inversion for mav disturbance rejection *Control Engineering Practice* **73** 79–90
[15] Oleynikova H, Taylor Z, Siegwart R and Nieto J 2018 Safe local exploration for replanning in cluttered unknown environments for microaerial vehicles *IEEE Robotics and Automation Letters* **3** 1474–1481
[16] Vitousek P M, Mooney H A, Lubchenco J and Melillo J M 1997 Human Domination of Earth's Ecosystems *Science* **277** 494–499
[17] Salmelin J, Pölönen I, Puupponen H H, Hämäläinen H, Karjalainen A K, Väisänen A and Vuori K M 2018 Hyperspectral imaging of macroinvertebrates—a pilot study for detecting metal contamination in aquatic ecosystems *Water, Air, & Soil Pollution* **229** 308
[18] Bacco M, Ferro E and Gotta A 2014 UAVs in WSNs for agricultural applications: An analysis of the two-ray radio propagation model *2014 IEEE SENSORS* pp 130–133
[19] Gago J, Douthe C, Coopman R, Gallego P, Ribas-Carbo M, Flexas J, Escalona J and Medrano H 2015 UAVs challenge to assess water stress for sustainable agriculture *Agricultural Water Management* **153** 9–19
[20] Manfredonia I, Stallo C, Ruggieri M, Massari G and Barbante S 2015 An early-warning aerospace system for relevant water bodies monitoring *2015 IEEE Metrology for Aerospace* pp 536–540
[21] Grenzdörffer G, Engelb A and Teichertc B 2008 The photogrammetric potential of low-cost UAVs in forestry and agriculture *The International Archives of the Photogrammetry Remote Sensing and Spatial Information Sciences* vol 37 pp 1207–1214
[22] Daponte P, De Vito L, Mazzilli G, Picariello F and Rapuano S 2017 A height measurement uncertainty model for archaeological surveys by aerial photogrammetry *Measurement* **98** 192–198
[23] Daponte P, De Vito L, Mazzilli G, Picariello F, Rapuano S and Riccio M 2015 Metrology for drone and drone for metrology: Measurement systems on small civilian drones *2015 IEEE Metrology for Aerospace* pp 306–311

AUTHOR INDEX